Springer Tracts in Civil Engineering

Springer Tracts in Civil Engineering (STCE) publishes the latest developments in Civil Engineering - quickly, informally and in top quality. The series scope includes monographs, professional books, graduate textbooks and edited volumes, as well as outstanding PhD theses. Its goal is to cover all the main branches of civil engineering, both theoretical and applied, including:

- Construction and Structural Mechanics
- Building Materials
- Concrete, Steel and Timber Structures
- Geotechnical Engineering
- Earthquake Engineering
- Coastal Engineering; Ocean and Offshore Engineering
- Hydraulics, Hydrology and Water Resources Engineering
- Environmental Engineering and Sustainability
- Structural Health and Monitoring
- Surveying and Geographical Information Systems
- Heating, Ventilation and Air Conditioning (HVAC)
- Transportation and Traffic
- Risk Analysis
- Safety and Security

Indexed by Scopus

To submit a proposal or request further information, please contact: Pierpaolo Riva at Pierpaolo.Riva@springer.com, or Li Shen at Li.Shen@springer.com

More information about this series at http://www.springer.com/series/15088

Efstratios Stylianidis

Photogrammetric Survey for the Recording and Documentation of Historic Buildings

Springer

Efstratios Stylianidis
Aristotle University of Thessaloniki
Thessaloniki, Greece

ISSN 2366-259X ISSN 2366-2603 (electronic)
Springer Tracts in Civil Engineering
ISBN 978-3-030-47309-9 ISBN 978-3-030-47310-5 (eBook)
https://doi.org/10.1007/978-3-030-47310-5

This Springer imprint is published by the registered company Springer Nature Switzerland AG
The registered company address is: Gewerbestrasse 11, 6330 Cham, Switzerland

The roots of education are bitter, but the fruit is sweet.

Aristotle

This book is dedicated to my parents,
Andreas and Stamatia ...

Preface

The seeds for this book were first planted in 2018, during my sabbatical leave in Columbia University and the Graduate School of Architecture, Planning and Preservation. At that time, apart from acting as Assistant Professor at the Aristotle University of Thessaloniki, I was serving CIPA-Heritage Documentation, the ICOMOS International Scientific Committee, as Secretary General. Since our last symposium, the 27th International Symposium (2019) held in Avila, Spain, where I have been elected to serve CIPA as the next President in a four-year term.

Image-based modelling techniques were increasingly being used in cultural heritage documentation to create image-based models of real-world objects. My decision to focus on the applications of photogrammetry to historic buildings to reinforce documentation seemed to resonate well with the international trends and the needs of preservation community.

This book reflects my 20+ years' experience in doing research in photogrammetry for the benefit of cultural heritage, mostly at the Aristotle University of Thessaloniki, but also through my deep involvement in CIPA. In pursuing my work, I have mostly absorbed in my work, on problems and solutions, techniques that are applicable in real-world conditions and that work well in practice in various disciplines. Therefore, this book has more emphasis on basic understanding of cultural heritage documentation and techniques that work under practical conditions.

Cultural heritage documentation is an interdisciplinary scientific field, and this book is suitable for studying and researching cultural heritage documentation, either originating from an engineering background or from arts and humanities.

In articulating and responding to cultural heritage documentation problems, I have often found it useful in several ways to draw inspiration from three high-level approaches:

- Philosophy: chase the "whys" as by documenting cultural heritage, we are assessing the values and significance of the heritage itself. Documentation is the DNA profile of cultural heritage (Fig. 1).

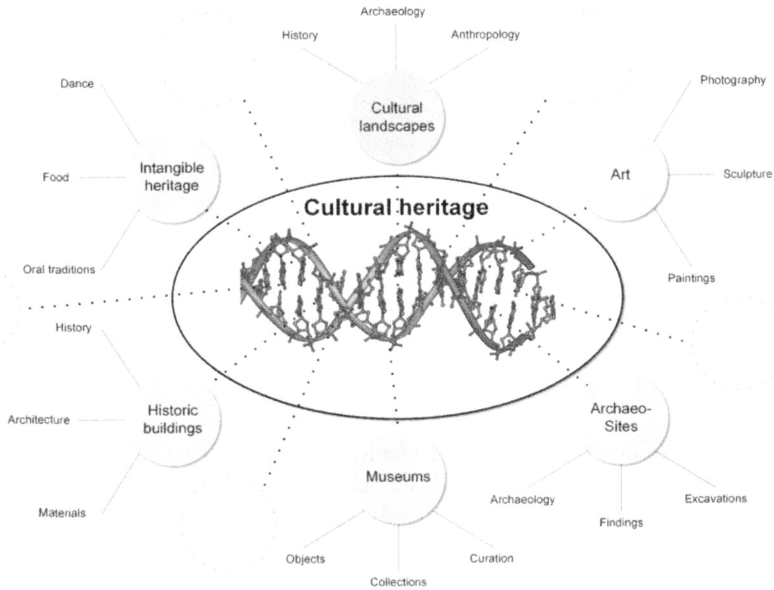

Fig. 1 DNA profile of cultural heritage

- Scientific: build detailed models of the historic structures to support preservation in an interdisciplinary context.
- Engineering: develop techniques that are simply describable and applicable but that are also known to work well in practice; test and understand their limitations.

These three approaches are intertwined, build on each other and are used throughout the entire book workflow. My personal research and development philosophy in this domain have a strong emphasis in the very endogenous features of cultural heritage. Those features who someone should dig more to discover, since cultural heritage is the soul of people and societies.

I would like to express my gratitude to Springer for the support and notably for the belief that this work would be delivered in a fair and sensible period of time to reach the respective community. I would also like to thank all my friends and colleagues for their technical input where needed and for the material provided and included in the book. Last but not least, I express my gratitude to my family for supporting me during my sabbatical leave, as well as while writing this book.

This book was written by one author. The main body of the book was prepared during my sabbatical leave while other parts after. I guess that the readers of this book will naturally find discontinuities of style, some repetition and some mistakes. I hope that all of you will excuse me for such faults. It is my own responsibility. I look forward to having your constructive feedback especially concerning those matters which should be rectified.

Thessaloniki, Greece Efstratios Stylianidis

Acknowledgements

All these years working in cultural heritage domain, I have been learning so many things all the time from my working practices. This is where my first acknowledgement must be: to those who have given me the opportunity to document a cultural heritage object. These are many and in different countries, in Cyprus, Georgia, Greece, Italy, Croatia, South Korea and many more.

Undoubtedly a book such as this depends on working experiences in cultural heritage documentation but also on information, verbal and written, formal and informal, collected from many and heterogeneous sources across the World. Inclusion in the Bibliography is an indirect form of acknowledgement. Occasionally sources have been used, nevertheless, the content has had to be reformed or modified properly to make it applicable to historic buildings and for this reason, cannot be directly acknowledged.

As the author of this book, I owe an excessive debt to many persons and friends across the world with whom I have had a professional interaction; however, it is impossible to nominate all such persons. I would very much like to acknowledge the support and plentiful advice I have received over many years from all my colleagues in the cultural heritage documentation international community. This amazing experience of collaboration by a documentation team of surveyors, architects, archaeologists, engineers and many more, working together, was encouraged by the urgent task of safeguarding cultural heritage.

The content of this book has been greatly polished through further experiences of lecturing at the Aristotle University of Thessaloniki, Faculty of Engineering, in Thessaloniki, Greece. I am grateful to my university for providing me this opportunity to educate young people over the years and do my research.

I owe a debt to all my friends in CIPA-Heritage Documentation for this incredible journey, since my first involvement many years ago. I served CIPA as a member of the Executive Board and Secretary General, and for the period 2020–2023, I will have the great honor to serve CIPA as the President. Since 2014, we are organizing a summer school (3D surveying and modelling in cultural heritage) in different places of the World.

I am very grateful to Columbia University and the Graduate School of Architecture, Planning and Preservation for hosting me cordially during my sabbatical leave in 2018. Most of this work has been realized during that time in those amazing libraries. I was working for many hours in such an inspiring place.

As usual, the author is fully responsible for what is written. Mistakes there may be, however, I hope they are not serious and that they will not delude anyone reading my book.

The book took a long time to be prepared and is dedicated to my parents for their valuable support and sacrifices in the cause of my career.

Contents

Symbols

x, y	Image point coordinates
X, Y, Z	Object point coordinates
x_o, y_o	Principle point coordinates
X_o, Y_o, Z_o	Camera projection center coordinates in the object coordinate system
c	Camera constant
λ	Scale factor
ω	Rotation about X-axis
φ	Rotation about Y-axis
κ	Rotation about Z-axis
s	Image scale
R	3D rotation matrix
ω, φ, κ	Parameters defining the image rotation with respect to the object coordinate system

Chapter 1
Introduction

Abstract The first chapter is an introduction to cultural heritage and the international context. Two international organizations, such as the United Nations and the European Union state culture and heritage, and through their fundamental principles and treaties unite these values. In addition to that, these organizations are playing a catalytic role over the years, regarding the protection of cultural heritage. The need for recording, documentation, conservation and preservation of cultural heritage, monuments and sites, was recognized even in the early times of human history. In general, the book focuses on the recording and documentation of cultural heritage, based on photogrammetry, which is one of the fundamental building blocks in preserving cultural heritage. The study of recording and documentation within the field of "data and information tool" is clearly placed within the historic preservation framework. Through the partnership of architects, conservationists, historians, engineers and other experts within an interdisciplinary context, many confusions and misconceptions may be avoided. Cooperation is a necessity. Occasionally, terminologies used are different and cause confusion, even while working in the same environment. There is always a need not to talk the same language uniformly, but to understand each other and respect the different methods, attitudes and backgrounds. Today, the radical changes in Information Communication Technologies and sensor technology are changing the landscape of data and the corresponding procedures.

1.1 Cultural Heritage International Context

Cultural heritage is the soul of the people and the societies; it is their asset but at the same time is set as their responsibility to future generations. Cultural heritage is rewarding humanity by providing classical and universal values. It provides the panhuman framework of mutual understanding, respect, liberty and expression.

Regardless if it is tangible or intangible, cultural heritage is an ecumenical treasure. It is the inheritance from the previous generations and the one for those to come. Cultural heritage is the lighthouse of humanity that enlightens the spirit, the souls and lives of the people. It is the origin of being mentally stimulated, to do or feel something great, notably to do a thing that is creative for them and the society too.

E. Stylianidis, *Photogrammetric Survey for the Recording and Documentation of Historic Buildings*, Springer Tracts in Civil Engineering,
https://doi.org/10.1007/978-3-030-47310-5_1

It is also a valuable resource to make people's lives better, improve their quality of life and contribute to economic growth, employment and social cohesion. Cultural heritage presentation and valorization are so important as preserving and valorizing cultural heritage is a game changer in defining those places as a spot where people can live, work, and visit.

Cultural heritage is for all and belongs to all. This is a diligent principle of humanity. However, sometimes it is susceptible to physical or human-made harm, to overuse or lack of funding and can end up in neglect, decomposition or in lethe. Belonging to all of us, cultural heritage implies our common responsibility to look after of it and preserve it.

Primarily, and for practical reasons, the protection of cultural heritage is a matter for local, regional and national authorities, depending the national legislation of each country. Nevertheless, cross-national organizations refer to culture and heritage, and through their fundamental principles and treaties consolidate these values. Two such examples are the United Nations (UN) and the European Union (EU).

The Charter of the United Nations (Fig. 1.1), also known as the UN Charter, was signed in San Francisco, United States, on 26 June 1945, by 51 countries representing all continents, as the foundational treaty of the UN, as an intergovernmental organization. The Charter led to the creation of the UN on 24 October 1945. Part of this Charter is the Statute of the International Court of Justice. The aim of the Charter is to save humanity from war, to reassure the human rights and the dignity and worth of every human person, to announce publicly the equal rights of men and women and of nations large and small, and to promote their well-being for all humankind. The Charter is the foundation of international peace and security (UN 1945).

Even though in the UN Charter the terms "cultural heritage" or "heritage" are absent, in different articles of the Charter, the term "culture" gets significant value. In Article 1.3 of the Charter, it is underlined that among the purposes of UN is "*to achieve international cooperation in solving international problems of an economic, social, cultural, or humanitarian character*". The Charter also emphasizes the international, economic and social cooperation, and in Article 55 it is stated that, "*UN shall promote solutions of international economic, social, health, and related problems; and international cultural and educational cooperation*".

EU is an evolving intergovernmental organization. Originally, the "*Treaty establishing the European Coal and Steel Community (ECSC)*" signed on 18 April 1951 (Figure 1.2a), aspired to create interdependence in coal and steel so that one country, of the founding countries, i.e. Belgium, France, Germany, Italy, Luxembourg, The Netherlands, could no longer mobilize its armed forces without others being aware. This eased mistrust and tensions after World War II (WWII).

The Treaties of the EU are a set of international treaties between the EU member states which defines the constitutional basis of the EU. The treaties establish the various EU institutions together with the functional framework. An agreement and ratification of every single party that has signed it is necessary for the amendment of the treaties.

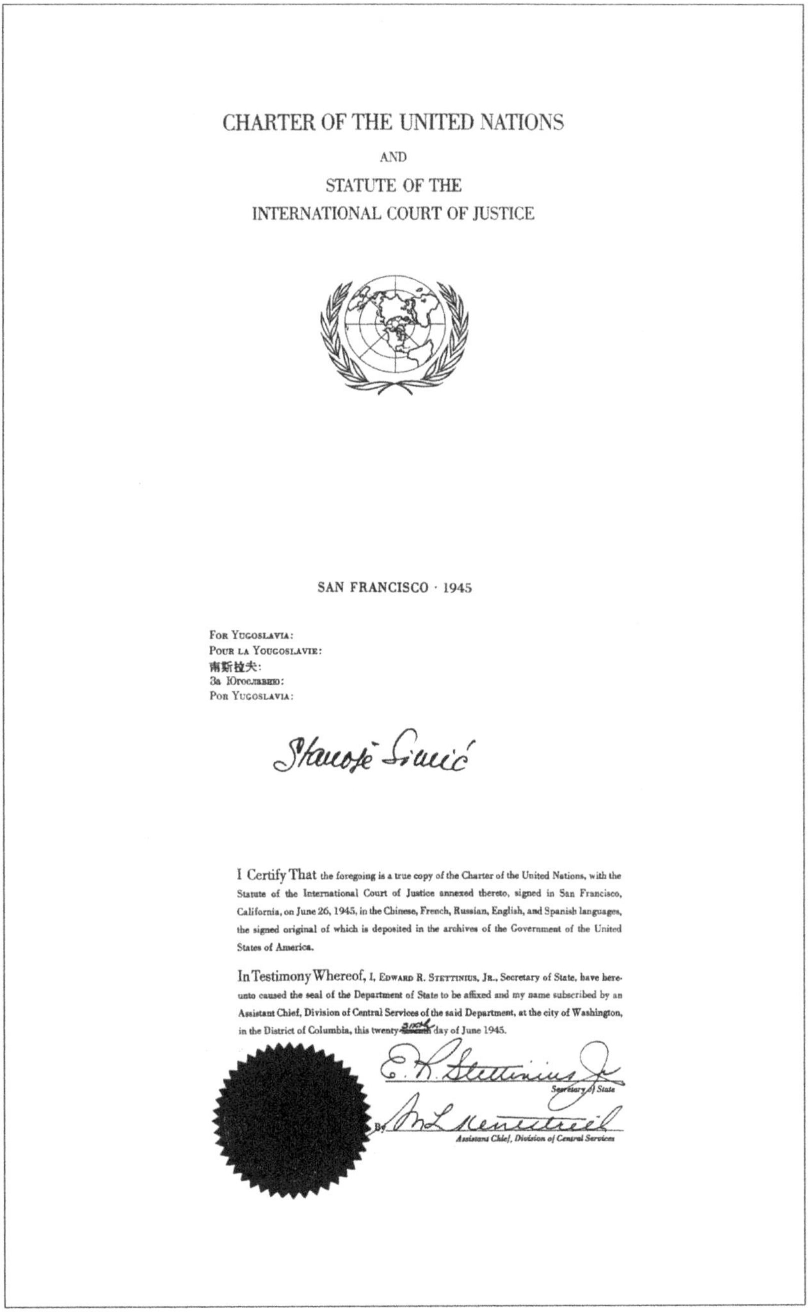
CHARTER OF THE UNITED NATIONS

AND

STATUTE OF THE

INTERNATIONAL COURT OF JUSTICE

SAN FRANCISCO · 1945

FOR YUGOSLAVIA:
POUR LA YOUGOSLAVIE:
南斯拉夫:
За Югославию:
POR YUGOSLAVIA:

I Certify That the foregoing is a true copy of the Charter of the United Nations, with the Statute of the International Court of Justice annexed thereto, signed in San Francisco, California, on June 26, 1945, in the Chinese, French, Russian, English, and Spanish languages, the signed original of which is deposited in the archives of the Government of the United States of America.

In Testimony Whereof, I, EDWARD R. STETTINIUS, JR., Secretary of State, have hereunto caused the seal of the Department of State to be affixed and my name subscribed by an Assistant Chief, Division of Central Services of the said Department, at the city of Washington, in the District of Columbia, this twenty-sixth day of June 1945.

Secretary of State

By

Assistant Chief, Division of Central Services

Fig. 1.1 Charter of the United Nations and Statute of the International Court of Justice. The first and the last page of the Chapter. (*Source* United Nations)

Fig. 1.2 18 April 1951 the Treaty of Paris establishes the European Coal and Steel Community 1.2(a); 25 March 1957 signing of the Treaty of Rome 1.2(b) and 7 February 1992 signing of the Maastricht Treaty 1.2(c). (*Source* European Commission)

Two core functional treaties:

1. the Treaty on the Functioning of the European Union, establishing the European Economic Community (EEC), was signed in Rome on 25 March 1957 (Fig. 1.2(b)), and entered into force on 1 January 1958, and
2. the Treaty on European Union, signed in Maastricht, The Netherlands in 1992 (Fig. 1.2(c)),

describe the operational framework of the EU. Over time, the treaties have been repeatedly amended.

The Treaties are handling the role, policies and the operation of the EU. Among the policies and internal actions is the European Social Fund which includes the cultural policy. The EU cultural heritage is a meaningful and diverse mosaic of cultural and creative expressions. It embodies natural, built and archaeological sites, monuments, museums, historic cities, and many more, including the knowledge, experiences, practices and traditions of citizens across Europe.

The European Commission (EC) has a specific role on the issue which is based on Article 3.3 of the Lisbon Treaty (EU 2007) which states: "*The Union shall respect its rich cultural and linguistic diversity, and [...] ensure that Europe's cultural heritage is safeguarded and enhanced*".

The Treaty on the Functioning of the EU gives the EC the explicit tasks of supporting the EU Member States towards the cultivation of culture, while respecting their diversity, and bringing "*[...] the common cultural heritage to the fore*".

1.2 Thematic Introduction

The need for recording, documentation, conservation and preservation of cultural heritage, monuments and sites, was recognized even in the early times of human history. The policies related to their protection, restoration and conservation have developed together with modernity, and now are recognized as a critical part of modern society's responsibilities. Since the 18th century, the goal of this protection has been defined as the cultural heritage of humanity; progressively, this has involved not only ancient monuments and historical works of art, but even entire regions (Jokilehto 1999). The development of attitudes has evolved over time, contributing to the establishment of an international society for world heritage protection and conservation during the 20th century. This ecosystem provided the adequate legal framework.

The scope of this study is not to touch on all issues related to the cultural heritage treatment domain. We focus on the recording and documentation of cultural heritage, which is one of the fundamental building blocks in preserving cultural heritage for the next generations. Even though the book will concentrate on historic buildings, it is easily recognizable that the majority of the content is applied to other "objects" or "places", such as archaeological sites.

A survey has many different meanings. In any case, it has a clear meaning to describe a scientific and professional study of a mapping process. We also have different application areas: i.e. topographical survey, archaeological survey, geological survey, etc. In this research, we emphasize in the application of three-dimensional (*3D*) surveying by means of photogrammetry, and we focus on the buildings as the objects of interest. In practice, it is a *3D* measured survey of buildings with special interest, such as historic buildings, in order to provide drawings in two-dimensional (*2D*) level or *3D* models and other survey outputs (e.g. measurements, images, etc.) of the building appearance, layout and construction.

1.3 Scientific Setting

The tools approach involves a set of different parameters that are very critical towards historic preservation, like ownership, regulation, incentives, property rights, and information (de Monchaux and Schuster 1997). All are very important, and none of these should be underestimated in this context.

Recording and documentation is set in the center of this study and thus, data and information is of primary interest. They are concerned in terms of acquisition, processing, storage and management; even exploitation could be added in this workflow. The value of data and information is priceless in historic preservation. Recording and documentation are considered an integral part of the "data and information tool" contributing to the preservation action.

The examination of recording and documentation within the field of "data and information tool" will be positioned explicitly within the context of historic preservation. By nature, historic preservation is an interdisciplinary domain, and no sole scientific system exists for the study of historic preservation. On the contrary, it involves several fields of art, science and technology. This is extremely useful and critical in order to perceive the nature of the physical and built environment. Historic buildings are almost everywhere, in small towns and big cities (Fig. 1.3). The related fields are meshed to avoid preconception of one-side hypotheses, i.e. to unilaterally support one (art) or the other side (science & technology). Besides, by enabling these fields to work together towards a common hypothesis, help the process of reaching an optimum conclusion, i.e. which are the optimum actions and measures to be taken for a historic building, so as to be preserved for the current and future generations?

Through the collaboration of architects, historians, engineers and other professionals with interdisciplinary background and training, many misunderstandings and misinterpretations may be avoided. Sometimes terminologies are different and get people confused, even while working in the same environment. There is always a need not to talk the same language uniformly, but to understand each other and respect the different approaches, attitudes and backgrounds.

RecorDIM was a five-year (2002–2007) international initiative among cultural heritage conservation organizations functioning together to bridge the existing gaps between the 'Information Users', such as specialists on conservation and the

Fig. 1.3 Historic buildings in small towns 1.3(a) and big cities 1.3(b). (*Source* Pexels—CC0 Creative Commons)

'Information Providers', such as photogrammetrists and surveyors (Letellier et al. 2007).

Nowadays, the radical changes in Information Communication Technologies (ICT) are changing the landscape of data and their respective processes, i.e. collection, management, analysis, visualization, and storage.

Big data, is such an example that it is able to change things we are doing. Big data is an expression that characterizes the large volume of data, structured and unstructured, that overwhelm almost everything nowadays. It is not the amount of data that is of great significance but what we do with the data that matters.

Machine learning as a field of artificial intelligence is another hot subject of conversation nowadays. In practice, it uses statistical techniques and procedures to give computer systems the ability to "learn" from data and improve their performance on a specific task.

There is a need for a new narrative. New technologies and learning procedures are stretching the gap between the different stakeholders, and the international organizations, the academic and research community should come closer to redefine the relationship between the different players in the preservation workflow.

1.4 Book Structure

The book is structured in seven different chapters, analyzing theoretical and practical issues for the topic.

In this chapter, '*Introduction*', the cultural heritage context in international level is discussed. The thematic and scientific setting of the book is also analyzed in this chapter.

In Chap. 2, '*Terminology and international framework for documentation*', the international organizations active in the area of cultural heritage documentation are presented. The various international charters, conventions and principles are providing the overall operational framework.

In Chap. 3, '*The need for documentation*', the international organizations active in the area of cultural heritage and related documentation and preservation activities are presented. The various international charters, conventions and principles are providing the overall operational framework and are part of this chapter. The human threats and natural hazards are the main factors for destruction of cultural heritage.

In Chap. 4, '*Historic buildings*', all the important features related to the historic buildings are analyzed, such as the architectural, historic and cultural interest, the location and sense of the buildings, the identity, the materials used for its construction, and the pathologies which appear over time. The role and use of historic building information systems is also discussed in this chapter.

The next three Chaps. (5, 6, 7) are presenting a typical workflow while working for the preservation of historic buildings. The first stage concerns the inspection tasks acting as the preliminary phase for project assignment to project understanding and planning. It is followed by the measurement activities by means of photogrammetry, concerning the condition of the building. Finally, the last stage is about the photogrammetric production workflow towards delivering various outcomes, necessary for the preservation of the historic buildings.

In Chap. 5, '*Planning: prior building surveying and documentation*', we discuss all the preparatory actions that are necessary prior to the recording and documentation phase. The project assignment, the user requirements, the technical specifications to be considered during the implementation, and the project planning are the most important features of this chapter.

Chapter 6, '*Measurements: introduction to Photogrammetry*', presents an introduction to photogrammetry and its principles. The various photogrammetric processes, such as image orientations and bundle adjustment are discussed within this chapter.

The last chapter of the book, Chap. 7, '*Production: generating photogrammetric outcomes*', is dedicated to the photogrammetric techniques leading to the mass production of *3D* points' coordinates. The dense point cloud and the production of the orthoimage as outcomes of the photogrammetric workflow are presented in this chapter.

In addition to the seven chapters, three additional Appendixes are used to present the three most figurative internationally recognized documents: i.e. "*The Venice Charter of 1964*", "*The World Heritage Convention of 1972*" and "*The ICOMOS Principles of 1996*".

References

de Monchaux J, Schuster, JM (1997) Five things to do. In: Schuster JM, de Monchaux J, Riley CA (eds.) Chap. 1 Preserving the built environment: tools for implementation, vol II. University Press of New England, Hannover and London

EU (2007) Treaty of lisbon amending the treaty on European Union and the treaty establishing the European community, signed at lisbon, 13 december 2007. https://eur-lex.europa.eu/legal-content/EN/TXT/PDF/?uri=OJ:C:2007:306:FULL&from=EN, Accessed by 17 Nov. 2018

Jokilehto J (1999) A history of architectural conservation. Oxford: Butterworth-Heinemann. p 354. ISBN: 07506-3793-5

Letellier R, Schmid W, LeBlanc F (2007) Recording, documentation, and information management for the conservation of heritage places. Guiding Principles. The Getty Conservation Institute. p 151. ISBN: 978-0-89236-925-6

UN (1945) Charter of the United Nations and statute of the international court of justice. San Fransisco. https://treaties.un.org/doc/publication/ctc/uncharter.pdf, Accessed by 27 Jan 2018

Chapter 2
Terminology and International Framework for Documentation

Abstract Terminology and the international framework for cultural heritage documentation is the theme of the second chapter. Although terminology used in 3D recording, documentation and preservation of historic buildings is very important, this study analyzes the main terms used in practice. The users should have a clear picture of each term before becoming involved in such projects. International organizations such as ICOMOS; the International Council on Monuments and Sites, ISPRS; the International Society for Photogrammetry and Remote Sensing and CIPA; CIPA—Heritage Documentation, one way or another, are working in the field of cultural heritage. The collaboration of ICOMOS and ISPRS produced CIPA. Over the last decades, various charters, conventions and principles have been adopted by several international organizations. Within this chapter, the most important of them, related to the recording and documentation of cultural heritage are discussed.

2.1 Glossary

Even though the terminology used in the *3D* recording, documentation and preservation of historic buildings is very important, this study does not get into details. We analyze the main terms used in practice. The users should have a clear picture of each term before becoming involved in such projects.

The bibliography on glossary and terminology is quite extensive, and it is originating from many and varying sources, conventions, charters, books, booklets, working documents, and many other important documents. It is also originating from different continents, cultural orientations, associations and authorities; national or international. Here in this book, our aim is to provide a summary of the most important terms used by deploying the relevant experiences. Whenever it was necessary to provide a term with an interpretation originating from a specific source, then this was done with appropriate reference.

For detailed and comprehensive analysis of the terminology used, further reading of the relevant bibliography is recommended. Especially, for the photogrammetric

E. Stylianidis, *Photogrammetric Survey for the Recording and Documentation of Historic Buildings*, Springer Tracts in Civil Engineering,
https://doi.org/10.1007/978-3-030-47310-5_2

terminology, Newby (2012) offered an expanded and updated edition of a "*Terminology Guide*" which was adopted as an official document of the International Society for Photogrammetry and Remote Sensing (ISPRS).

2D: something expressed in two-dimensions (*2D*) and it is usually characterized by the Cartesian *X, Y* coordinates.

3D: something expressed in three-dimensions (*3D*) and it is usually characterized by the Cartesian *X, Y, Z* coordinates.

3D modelling: the process of creating a *3D* computer model of an object, e.g. a building, that precisely reproduces the object form as it is in reality.

3D surveying: the technique of determining the terrestrial *3D* positions of points in an object by using different types of measurements, e.g. topographic, photogrammetric, laser scanning, etc.

Accuracy: refers to the degree to which the results of a recording activity for an object comply to the object metric value. In practice, it is highly connected to the scale and the precision of the surveying technique used, as well as to the quality of the object graphic record.

Addition: this is a new construction added to an existing building or structure, e.g for aesthetic reasons.

As-built: describes how a structure was constructed, compared to the initial design.

As-is: provides the picture of the present condition of a building, structure, etc.

Building: a building, such as a house, church, or a similar construction is built mainly to host human activities. It may also be used to refer to a historically and functionally related unit, e.g. church, library, theater, etc.

Building Information Modelling (BIM): a smart *3D* model-based process that allows architecture, engineering, and construction (AEC) specialists to gain an accurate and deep intuitive understanding and tools to more professionally plan, design, construct, and manage buildings.

Computer-aided design (CAD): concerns the use of computer systems in order to assist the creation, modification, analysis, or optimization of a drawing which is typically in *2D* or *3D*.

Conservation: encompasses all the processes for taking care of a cultural heritage object so as to keep possession of its historic and cultural significance. The purpose of *conservation* is to study, record, maintain and restore the important cultural assets of the historic buildings. From the practical point of view, *conservation* includes a selected package of tasks, such as inspection, recording, documentation, preventive conservation, treatment, restoration, preservation and reconstruction.

Control Points (CPs): refers to the points of known position that determine a coordinate system, in which all other measurements can be referenced. If the points are on the ground/terrain, the term Ground Control Points (GCPs) is usually used.

Coordinate system: any system that permits the use of numeric values, e.g. in *3D—X, Y, Z* coordinates, to uniquely determine the location of a point on the Earth surface, or any other object (e.g. a building).

Culture heritage: according to the UNESCO World Heritage Convention 1972 (Sect. 2.3.3) (Annex B), Article 1, the following are considered as *cultural heritage*:

- *"monuments: architectural works, works of monumental sculpture and painting, elements or structures of an archaeological nature, inscriptions, cave dwellings and combinations of features, which are of outstanding universal value from the point of view of history, art or science;*
- *groups of buildings: groups of separate or connected buildings which, because of their architecture, their homogeneity or their place in the landscape, are of outstanding universal value from the point of view of history, art or science;*
- *sites: works of man or the combined works of nature and man, and areas including archaeological sites which are of outstanding universal value from the historical, aesthetic, ethnological or anthropological point of view."*

Deformation monitoring: refers to the organized and typical measurement and monitoring of the change in shape or dimension of an object, e.g. a building, as a consequence of the action of putting pressure (natural or human-made) to it.

Digital Elevation Model (DEM): it is the elevation model of a surface in digital mode which is expressed in *3D* coordinates (*X, Y, Z*) and in a digital format.

Demolition: action or process that destroys partly or wholly a building or structure.

Documentation: an activity for the systematic collection and archiving of historic building records, in order to preserve them for future use. The entire collection of records, written and graphic are taken up during the investigation and treatment of the building.

Digital Surface Model (DSM): it is the digital model of a surface that fully envelopes all objects in the space. In the case the object is the surface of the Earth, then the surface which is expressed in *3D* coordinates (*X, Y, Z*) in a digital format, includes vegetation, man-made objects, etc.

Digital Terrain Model (DTM): refers to the digital elevation model of the terrain surface (only), expressed in *3D* coordinates (*X, Y, Z*) in a digital format.

Enlargement: extend a building or structure beyond its present outline, commonly through the construction of an additional or new external feature.

Fabric: all the physical materials of the building. *Fabric* includes building interiors and sub-surface remains.

Façade: the exterior building view that is the architectural front, occasionally notable from the other faces, having architectural and/or impressive decoration.

Geographic Information System (GIS): a computer system able for capturing, storing, managing, analyzing, and presenting spatial or geographic data; i.e. data identified according to location.

Global Navigation Satellite System (GNSS): a standard generic term for satellite positioning and navigation systems that provide autonomous geospatial positioning with worldwide coverage. Common GNSS are GPS (US), GLONASS (Russia), Beidou (China) and Galileo (EU).

Graphic record: the collection of drawings, rectified images, ortho-photomosaics, *3D* models, able to describe the physical configuration of a historic building with its dimensional and architectural characteristics.

Hardware: in ICT, it is the physical part of computers, telecommunications, and other respective devices. As a collective term, it includes not only the computer, but also the cables, power supply, and peripheral devices such as the keyboard, mouse, printers, audio speakers, etc.

Historic object: an object that has an artistic and small-scale construction, included in or qualified for designation, and which is significant due to its connection with a historic event, movement, place or person.

Historic monument: according to the Organization for Economic Co-operation and Development (OECD), is a "*fix asset that is identifiable because of particular historic, national, regional, local, religious or symbolic significance; it is usually accessible to the general public, and visitors are often charged for admission to the monuments or their vicinity*" (OECD 2001).

Historic building: a building that is momentous in the history of architecture, that integrates important architectural structures, or that played major historic roles in local, regional, national or international cultural or social development. A historic building may be or not, officially designated by the accredited authorities.

Historic preservation: according to the United States National Historic Preservation Act (NHPA 1966), "*includes identification, evaluation, recordation, documentation, curation, acquisition, protection, management, rehabilitation, restoration, stabilization, maintenance, research, interpretation, conservation, and education and training regarding the foregoing activities or any combination of the foregoing activities*".

Information management: concerns the organizational activity that includes a sort of steps such as information acquisition from one or more sources (heterogeneous information), its distribution to those interested in, and its archiving in the proper means. It contains all the general management phases of planning, organizing, structuring, processing, evaluating and reporting.

Interpretation: all the various ways of presenting the cultural significance of a historic building.

Intervention: an action applied by someone, excluding demolition or destruction, that outcomes to a change to a part of a historic building.

Laser scanning: refers to the process of using a *3D* laser scanner, i.e. a device that collects *3D* coordinates of an object surface in an automatic and systematic pattern. The result is usually expressed by a point cloud in terms of (X, Y, Z) coordinates, with a massive number, and in (near) real time mode.

Management plan: a strategic document that describes the way to care for the cultural heritage property. This plan foresees rather further than a conservation plan by incorporating other important parameters affecting the cultural heritage property use, such as social, political and economical context.

Materials: any material or combination of materials used for the construction or reconstruction of a historic object.

Measured survey: refers to the activity of generating drawings by any type of measurements, e.g. hand, tape, total stations, photogrammetry, laser scanning, etc.

Mesh: a polygonal subdivision of the geometric model surface, that it is also considered as a triangulated model or a polygon model.

Metadata: it is the data used to describe other data or provide information about one or more data aspects. It is an absolutely necessary component of the data management process, as it is used to sum up basic information about data in a way that can facilitate data tracking and working with specific data.

Monitoring: frequent and repeated measurements of changes in order to analyze and evaluate changes occurring on a historic building.

Monuments: according to the UNESCO World Heritage Convention 1972 (Sect. 2.3.3) (Annex B), Article 1, "*architectural works, works of monumental sculpture and painting, elements or structures of an archaeological nature, inscriptions, cave dwellings and combinations of features, which are of outstanding universal value from the point of view of history, art or science*".

Object: it is used to differentiate from buildings and structures the construction that is mainly creative in nature or rather small in scale. Even though an *object* may be considered as movable, by nature or design, it is connected with an explicit setting or environment, e.g. a sculpture.

Open-source software (OSS): refers to the software that its source code can be inspected, modified, and enhanced by any person. The authors of an OSS make the source code available to other persons who would like to view the code, copy it, learn from it, change it, or share it. According to the Open Source Initiative, the distribution terms of an OSS must comply with specific criteria (OSI 2007).

Orthoimage: is an image that is geometrically corrected, the so-called "orthorectified" image. The scale is uniform and unlike an uncorrected image, an orthoimage (alternatively an orthophoto or orthophotograph), can be used to measure real distances, because the image has been adjusted properly for elevation, lens distortion, and camera tilt.

Photogrammetry: the art, science, and technology that contains the methods of image measurement and interpretation in order to obtain the location and the shape of any type of object, usually by using coordinates, from one (*2D*) or more photographs/images (*3D*) illustrating this object.

Point cloud: a set of data points in a *3D* coordinate system, which are defined by using (*X, Y, Z*) coordinates, and it is able to represent the captured surface of an object or a historic building.

Restoration: means the process of returning a historic building to a known previous condition by removing accretions or by gathering together again existing features without using any new material.

Preservation: means all the proper processes that are necessary to be applied in order to maintain a historic building in its existing condition and, further than that, in decelerating deterioration.

Preventive conservation: the activity targeting a historic building, that embraces all measures, procedures and actions aimed at dodging and minimizing future deterioration or loss, i.e. mitigating deterioration, decay, and damage to cultural heritage property. The measures and actions do not intrude the materials and structures, and they do not revise the property appearance. The activity usually refers to lighting, air quality, integrated pest management, fire protection, emergency preparedness and response, etc.

Protection: refers to the action of designing and applying measures that are able to defend the historic building condition and safeguard it from deterioration, damage, decay, loss or attack. In most of the cases, protection is considered a treatment of a provisional nature for the cultural heritage and anticipates future preservation and conservation.

Reconstruction: means the process of returning a historic building to a known previous condition, and it is distinguished from restoration by the action of introducing new materials.

Recording: means new information acquisition originating from all the activities on a historic building, containing recording, study and analysis, conservation, use, monitoring and management.

Records: refers to the outcomes of recording process.

Rehabilitation: according to the U.S. Department of the Interior, National Park Service, Technical Preservation Services, "*rehabilitation is defined as the act or process of making possible a compatible use for a property through repair, alterations, and additions while preserving those portions or features which convey its historical, cultural, or architectural values*" (NPSc 2017).

Renovation: a broad term to depict the modification process followed for a historic building in order to prolong its useful life. Besides, it is used to feature the improvements made to existing historic buildings. In practice, *renovation* may contain rehabilitation and several other activities.

Repair: this term refers to the replacement or correction of broken, damaged or not working features of a historic building, either in the inner part or outside.

Restoration: according to The Venice Charter—1964 (Article 9), "*the process of restoration is a highly specialized operation. Its aim is to preserve and reveal the aesthetic and historic value of the monument and is based on respect for original material and authentic documents. It must stop at the point where conjecture begins, and in this case moreover any extra work which is indispensable must be distinct from the architectural composition and must bear a contemporary stamp. The restoration in any case must be preceded and followed by an archaeological and historical study of the monument*" (Sect. 2.3.2) (Annex A).

Scale: refers to the relationship or ratio of a size (e.g. a distance) on the drawing and the actual physical size in reality (historic building). A large scale means higher accuracy and higher quality detail while small scale reflects the opposite.

Sketch diagram: analytical freehand drawing that illustrates the key relationships between components and allows understanding the historic building and supporting its survey.

Software: it is a generic term for the different sorts of programs used to operate computers and other relevant devices. It can be considered as a variable component of the computer, if hardware is the invariable. Many times, it is split into the application software, i.e. the programs that the users are employing, and system software, e.g. the operating system, for example Linux or Windows.

Source code: it is the part of the software that most computer users do not ever see. In fact, it is the version of the software as it is originally written by a programmer, in plain text, i.e., in human readable alphanumeric characters.

Stabilization: the action or process of intervening that can be used as a temporary step on a seriously deteriorated historic building, or it may contain the long-term structure unification.

Structure: refers to the functional constructions made usually for purposes excluding building human shelter, such as bridge, lighthouse, tunnel, windmill, etc.

Style: a type of architecture distinguished by structure, distinct characteristics and decoration. Many times, it is linked in time scale.

Triangulated Irregular Network (TIN): it is a representation of a surface organized in a network of non-overlapping triangles.

Total station: a standard survey device that it is designed for measuring distances, horizontal and vertical angles in topographic and geodetic works. The onboard computer performs trigonometric calculations by combining the angles (horizontal and vertical) with distances to determine (X, Y, Z) coordinates.

Use: means the functional mode of a historic building, along with the activities and practices that possibly will occur at the historic building.

2.2 International Organizations

2.2.1 International Council on Monuments and Sites—ICOMOS

ICOMOS is an international, non-governmental organization working for the conservation and protection of cultural heritage, the World's monuments and sites. ICOMOS is the principal advisor of UNESCO in matters related to the conservation and protection of monuments and sites. ICOMOS has an internationally recognized role under the World Heritage Convention (Sect. 2.3.3) (Annex B) to offer suggestions to the World Heritage Committee and UNESCO on the nomination of new sites to the World Heritage List. It is devoted to actively encouraging the application of theories, methodologies, and scientific methods, techniques and tools to the conservation of the archaeological and architectural heritage (Fig. 2.1).

The founding principles of ICOMOS stand on the International Charter on the Conservation and Restoration of Monuments and Sites (the Venice Charter) of 1964 (Sect. 2.3.2) (Annex A). As an international organization has members which are typically professionals and experts, constituting a multidisciplinary network of anthropologists, archaeologists, architects, art historians, engineers, geographers, and town planners. These members offer their knowledge and expertise to improving heritage

Fig. 2.1 The logo of ICOMOS

Fig. 2.2 The logo of ISPRS

preservation, the standards, techniques and tools for different categories of cultural heritage property, such as archaeological sites, buildings, cultural landscapes, and historic cities. ICOMOS was founded in 1965 in Warsaw as a result of the international adoption of the Venice Charter the year before (ICOMOS 2018).

2.2.2 *International Society for Photogrammetry and Remote Sensing—ISPRS*

"*Photogrammetry and Remote Sensing is the art, science, and technology of obtaining reliable information from non-contact imaging and other sensor systems about the Earth and its environment, and other physical objects and processes through recording, measuring, analyzing and representation*". ISPRS is an international, non-governmental organization devoted to the development of international cooperation, in scientific and professional level, for promoting photogrammetry, remote sensing, and spatial information sciences, in terms of theory, methodology, and scientific techniques, tools and applications. It was established in 1910 as the International Society for Photogrammetry (ISP), and 70 years later (1980), the Society changed its name to the International Society for Photogrammetry and Remote Sensing. ISPRS is the oldest international organization in the fields of photogrammetry, remote sensing and spatial information sciences and achieves its aims by (ISPRS 2018) (Fig. 2.2):

- Supporting research and development, scientific networking and inter-disciplinary activities.
- Accommodating education and training activities with emphasis in less developed countries.
- Increasing public recognition of the contributions of its scientific fields for the benefit of humankind and the sustainability of the environment.

Fig. 2.3 The logo of CIPA

2.2.3 *CIPA Heritage Documentation—CIPA*

CIPA (Fig. 2.3) is one of the first ICOMOS Scientific Committees that were established after the foundation of ICOMOS. According to Patias (2004), all began on 4–6 July 1968 in Saint Mandé/Paris, France, in a Colloquium on the applications of photogrammetry to architecture organized by ICOMOS and Maurice Carbonnell (CIPA's First and Honorary President). It was created in cooperation with ISPRS, to assist conservation professionals and experts with activities of cultural heritage recording and documentation. Originally, it was founded as the Comité International de la Photogrammétrie Architecturale (English: "International Committee of Architectural Photogrammetry"). According to Waldhäusl (2004), Hans Foramitti is considered the co-founder of CIPA. Gradually, this Committee evolved into a multidisciplinary organization that uses technologies for measurements, data management, visualization and representation for the benefit of recording, documentation and conservation of cultural heritage. For this reason, its original name was substituted by CIPA Heritage Documentation or simply CIPA. Primarily, CIPA had the work to develop recording tools to fulfil the documentation principles explicitly written in Article 16 of the Venice Charter. Nowadays, CIPA is a dynamic international and multidisciplinary organization that keeps up with technology and ensures its usefulness for cultural heritage conservation, education and dissemination (CIPA 2019).

2.3 Charters—Conventions—Principles

The notion of "global values" of the world cultural heritage, which was progressively developed during the 19th century, was eventually to touch a recognized expression on international level. The acknowledgment of cultural heritage, versatile values and its increasing and decisive contribution to understanding of humankind development are the outcome of an international approach that was introduced during the 20th century.

A review of the key international documents in the field of world cultural heritage recording, documentation, conservation and preservation delivers a consistent basis for understanding the origin of an interdisciplinary approach towards preservation and conservation philosophy on local, national, regional and international levels. It is not the scope of this book to present all the international framework of charters, conventions or any other important (national or international) initiative for the cultural heritage preservation. Essentially, this is not possible, and, in any case, is not the main focus of this book. The attempt is to provide the most important milestones and especially those critical initiatives that placed survey, recording and documentation as an integral part in the conservation and preservation process.

As most of the ICOMOS-related cited documents, either charters, or principles, or conventions, or other international standards can be found on journal and conference publications or around the Internet, a detailed analysis of all will not take place. The author is aware that this overview to the charters, conventions and principles is not complete, however, this is not the target of this book. The human values are the driving-force behind conservation and preservation of cultural heritage. Conservation and preservation of cultural heritage cannot be considered without recording and documentation.

The 6th International Congress of Architects of 1904 in Madrid (Locke 1904) provided a brief draft of recommendations concerning preservation and restoration of architectural monuments. The brief recommendations established the first attempt to introduce architectural conservation principles. Amongst others, these brief recommendations highlighted:

- The significance of negligible intervention in taking care of ruined structures.
- The finding a functional use for historic buildings.
- The principle of style unity, which encourages restoration according to a single stylistic expression.
- That "*restoration should be effected in the original style of the monument, so that it may preserve its unity, unity of style being also one of the bases of beauty in architecture, and primitive geometrical forms being perfectly reproducible*".
- The division of monuments into two groups: "`dead` *monuments, i.e. those belonging to a past civilization or serving obsolete purposes, and* `living` *monuments, i.e. those which continue to serve the purposes for which they were originally intended*".

World War I (WWI) led to the destruction of many cultural heritage monuments. This development encouraged the international community to create the International Commission for Intellectual Cooperation. The League of Nations Commission for Intellectual Cooperation was founded in 1922 as an advisory organization for the League of Nations aimed to encourage worldwide exchange between scientists, researchers, teachers, artists and intellectuals. A few years later, in 1926, the International Commission for Intellectual Cooperation, League of Nations, creates the International Museums Office (IMO). In 1936, IMO carried out a study, which would result in a draft Convention for the Protection of Historic Buildings, and Works of Art in Times of War presented to the League of Nations' Council and General Assembly in 1938. In 1946, with the establishment of the United Nations (UN) system, UNESCO and the International Council of Museums (ICOM), the International Commission for Intellectual Cooperation, League of Nations, ceased its function (Daifuku 1998).

2.3.1 Athens Charter of 1931

The "First International Congress of Architects and Technicians of Historic Monuments" on the protection and conservation of cultural and historical monuments, took place in Athens 1931. The resolutions of this congress were issued as the "*Athens Charter for the Restoration of Historic Monuments*", or simply known as the "*Athens Charter*". Indeed, this was the first international document recognized at an interdisciplinary level and dealt with general notions and principles related to protection, conservation and restoration.

The Athens Charter general trend was to service, where possible, conservation instead of restoration with respect to historic monuments' authenticity. It also advised to reinforce international cooperation in technical matters, as well as form an international center that would assemble the documentation on world cultural heritage.

The Athens Charter (1931) is often confused with the conclusions of the IVth International Congress of Modern Architects held, also in Athens, in 1933. The resolutions of this new congress led to the urbanism charter and share the same name apart from the calendar year. This document paid particular attention on the principles of modern urban planning, despite the fact that it had some recommendations on interventions within old urban fabric. Via a parallel reading of the two texts, the circumstances of their origin were traced and show how innovative and original the ideas advanced were (Iamandi 1997).

2.3.2 Venice Charter of 1964

As already explained in Sect. 1.1, the formation of the UN in San Francisco in 1945 led to changes to the principles of cultural heritage protection and conservation, and also to the creation of UNESCO in London, in 1945. As a consequence, three international organizations were created, namely the:

- International Council of Museums—ICOM in 1946
- International Union for Conservation of Nature—IUCN in 1948
- International Conservation Center for Restoration of Monuments—ICCROM in 1959

In 1964, the 2nd International Congress of Architects and Technicians of Historic Monuments was organized in Venice. Definitely, it is considered to be a milestone international event. The resolutions included the International Charter for the Conservation and Restoration of Monuments and Sites, simply known as *The Venice Charter* (Annex A), which became a fundamental international 'Bible' on conservation and restoration theory and practice. Apart from this, one of the resolutions put forward the creation of an organization that would coordinate the international efforts for the preservation and evaluation of the world cultural heritage. In 1965, UNESCO established the International Council on Monuments and Sites (ICOMOS) (Sect. 2.2.1), which has accepted the Venice Charter as its ethical principle.

The Venice Charter intoned the necessity to respect the authenticity of the historic monuments while their use should be the appropriate one, as marked in Article 3, i.e. "*The intention in conserving and restoring monuments is to safeguard them no less as works of art than as historical evidence*". It is considered as the first paper that connected the historic monument to its surroundings. As written in the Article 7, the matter of interest was also specified to the protection of the setting. The monument is considered "*inseparable from the history to which it bears witness and from the setting in which it occurs*".

There are also two important Articles in the Venice Charter, namely Article 2 and Article 16, that mark the contribution of science and techniques to be used for studying and precise recording and documentation of cultural heritage. Especially, Article 16 of the Venice Charter, which describes exactly the nature of practice standards for monuments of great significance or elements of monuments, and archaeological sites.

"*Article 2: The conservation and restoration of monuments must have recourse to all the sciences and techniques which can contribute to the study and safeguarding of the architectural heritage.*"

"*Article 16: In all works of preservation, restoration or excavation, there should always be precise documentation in the form of analytical and critical reports, illustrated with drawings and photographs. Every stage of the work of clearing, consolidation, rearrangement and integration, as well as technical and formal features identified during the course of the work, should be included. This record should be placed in the archives of a public institution and made available to research workers. It is recommended that the report should be published.*"

The Venice Charter has been the reference point for principles governing architectural restoration and conservation for decades. The Venice Charter principles have also been generally recognized as the primary policy guidelines for the assessment of cultural heritage sites on UNESCO's World Heritage List. Jukka Jokilehto, ex. Head of the Architectural Conservation programme at ICCROM prepared an overview of the context and the history of the Venice Charter (Jokilehto 1998).

2.3.3 World Heritage Convention of 1972

The concept of founding an international movement for protecting heritage moved out after WWI. The “UNESCO Convention Concerning the Protection of the World Cultural and Natural Heritage” of 1972 (Fig. 2.4), as it is originally known, elaborated from the uniting of two separating movements. One focusing on the preservation of cultural sites and one targeting the conservation of nature. According to UNESCO (UNESCO 1972), in 1959, the event to build Aswan High Dam in Egypt launched an international campaign to safeguard the Abu Simbel and Philae temples. Similar campaigns were organized in other countries (Italy, Pakistan, Indonesia) to safeguard outstanding cultural sites. Driven by these events, UNESCO, with the support of ICOMOS, drafted the convention on the protection of cultural heritage. In 1965, a White House Conference in Washington, D.C. called for a “World Heritage Trust” that would raise levels of international cooperation to protect the natural areas and historic sites. In 1968, the IUCN elaborated similar proposals for its members. In 1972, these proposals were introduced to the UN Conference on Human Environment in Stockholm. In the end, a single text was agreed upon by all parties concerned, and on 16 November 1972, the “Convention Concerning the Protection of World Cultural and Natural Heritage” (Annex B) was adopted by the UNESCO General Conference.

Fig. 2.4 Signature of the World Heritage Convention by René Maheu, UNESCO Director-General, 23/11/1972. (*Source* UNESCO—CC-BY-SA 3.0)

In 2011, almost 40 years after the introduction of the UNESCO World Heritage Convention, Cleere (2004) pointed out that even though the convention ought to be rather the most effective way of guaranteeing the continuity state of the past for the benefit of future generations, he is worried that the member states of this convention, and UNESCO as well, have lost their way from the original objectives of the convention. He believes that an independent survey should take place to present what has been achieved and what has gone wrong. Cleere is criticizing issues such as the growth of the World Heritage List (originally the List would not exceed one hundred heritage properties, with equal numbers of them in cultural and natural), the "World Heritage Competition" among a number of member states which believe that they have as many as possible of the heritage properties from all of human history that are found on its territory inscribed on the list, the implications of inscription in the World Heritage List as the funding from the World Heritage Fund is severely limited to cover the huge number of inscribed heritage properties, etc. He finally argues that if the convention is to continue to exist and to have the quality of being logical, a thorough examination and reorientation should be assumed, in the points of criticism.

In the fortieth celebration of the World Heritage Convention in 2012, Rodwell (2012) prepared an article presenting the author's thoughts on the achievements and future directions for the Convention, focalizing to a great extent on cities. The author notes that Convention cannot be changed without prior individual agreement of all signatory parties. However, the wording used 40 years back, has proved resilient to the passage of time. Rodwell is also focusing on the world heritage brand. He argues that even though there are the innumerable assertive benefits of the World Heritage brand, there are examples of negative impacts that are a reason for concern. He mentions two examples: the one of the city of Zamość in Poland and the city of Xi'an in China. He is closing his article by underlining that the Convention "*is at serious risk of losing this cultural and environmental ethos in favor of development that is predicated primarily in terms of selective economic gain, often to the prejudice of local communities.*"

In an attempt to evaluate the implementation of the Convention, Vigneron (2016) performed a study in 10 countries, namely Australia, China, France, Germany, Italy, Japan, Spain, Switzerland, the United Kingdom and the United States, which were part of a research network funded by the UK's Arts and Humanities Research Council in 2012–2015. The target of this study was to carry out a survey to investigate the selection process of cultural properties at the national level. That survey was based on a questionnaire that targeted the identification of national practices with respect to the recognition and candidature of sites for suggestion on the national Tentative List and then in the World Heritage List. Even though the World Heritage Committee has introduced the nomination process, the participating countries to the Convention each have their own way of carrying out the nomination. The survey in the 10 countries has shown that the apparent integrated process still permits the countries to act differently in the different process steps.

In partnership with Oral Archives Initiative of UNESCO, Cameron and Rössler (2011) attempted to record the historical testimonies of the protagonists (voices of the pioneers) and those who played an important role in the creation and introduction of the World Heritage Convention. The authors interviewed the active participants of that period, to complement the existing literature and bulky documentation from the significant meetings of that era. The uniqueness of this research project stands on the significant role that the "voices of the pioneers" will play in lighting up that period. The authors interviewed 31 persons.

The World Heritage Committee is the main body in charge of the World Heritage Convention implementation. It is responsible for defining the use the World Heritage Fund, i.e. the fund that was established to support and distribute the financial assistance upon requests from the participating countries. Since 1977, the Committee has elaborated precise criteria for the inscription of properties on the World Heritage List. These are all contained in a paper entitled "*Operational Guidelines for the Implementation of the World Heritage Convention*" (UNESCO 2017). These guidelines are revised by the Committee from time to time to reflect new meanings, knowledge and practical experiences.

2.3.4 The Burra Charter of 1979, 1999 and 2013

In 1979, the Australian ICOMOS adopted "*The Charter for the Conservation of Places of Cultural Significance*" at a meeting which took place at the historic mining town of Burra in South Australia. Since then, it is very well known as the "*The Burra Charter*". The Charter accepted the philosophy and concepts of the ICOMOS Venice Charter (Sect. 2.3.2) but adapted in practical and useful way for Australia. The Charter was revised in 1999. Before that, minor revisions were made in 1981 and 1988. The Charter was adopted in its current version by the Australia ICOMOS in 2013 (ICOMOS 2013).

The Burra Charter defines the basic principles and procedures, and the nationally (Australia) accepted standard, to be followed in the heritage conservation. The proposed heritage conservation framework can be applied to a monument, historic building, archaeological site, or any other heritage object, structure or even a whole region. It does not propose the methods and techniques to be used or the manner in which cultural heritage should be looked after.

According to the Burra Charter, conservation of cultural heritage is a process that is composed of three stages:

- Understanding the significance.
- Developing policy.
- Managing in accordance with the policy.

The requisite of any cultural heritage conservation project is understanding the object and collecting data about its normal status prior to any process and intervention that might lead the object to become different. Besides, cultural heritage is threatened by natural and human-made disasters, ageing, etc., and thus, no one can guarantee their perpetuity. This is why documentation is an important and imperative process (Hassani 2015).

According to the Burra Charter, people involved in the conservation of the cultural heritage places should:

- Be savvy and care of the cultural significance of the place, before taking any decisions about its future, since conservation is for present and future generations. Any decision should be in accordance with the principle of inter-generational equity.
- Protect the setting of the place, including the visual and sensory setting but also the retention of spiritual and other cultural relationships that contribute to its cultural significance.
- Interpret and present the place in a way appropriate to its significance while at the same time reinforcing understanding and engagement.
- Involve and create opportunities for the participation of the communities and other cultural groups that are associated with the place.
- Make use of all the knowledge, skills and disciplines which can contribute to the study and care of the place.
- Provide security for the place.
- Provide an appropriate use.

Especially, for the documentation aspect, Burra Charter refers to the records of the conservation of a place that should be archived and protected properly "*in a permanent archive and made publicly available, subject to requirements of security and privacy, and where this is culturally appropriate*".

The Burra Charter is adopted by an ICOMOS National Committee, namely Australia, and not by the General Assembly of ICOMOS (ICOMOS 2018).

2.3.5 ICOMOS—Principles for the Recording of Monuments, Groups of Buildings and Sites of 1996

As Letellier et al. (2007) indicates in his publication, this initiative started one year earlier the official ratification, in 1995, during an ICOMOS international meeting in Kraków, Poland. Letellier was authorized to organize an ad hoc group of experts to review an ICOMOS UK and ICOMOS France document on recording principles. Fifteen experts from UNESCO, ICOMOS, ICCROM and CIPA participated in this group and the concluded to a revised document, titled "Principles for the Recording of Monuments, Groups of Buildings and Sites". The integrated documented was presented and adopted during the 11th ICOMOS General Assembly in Sofia, Bulgaria, in 5–9 of October 1996, as given in Annex C.

The ICOMOS document, attains for the first time since the founding of ICOMOS and the Venice Charter in 1964 (Annex A), to specialize the matters of recording cultural heritage. It manages to define a framework of recording principles in which cultural heritage is defined, conserved and managed. The document presents the principles in an uncomplicated and easy to understand way, and it is structured in five parts.

- **Preamble**
  This is the introductory part of the document that establishes the why of recording and refers to Art. 16 of the Venice Charter (Annex A), which requires responsible institutions and (expert) individuals to record the nature of the cultural heritage. Three of the most important terms in the documentation and conservation process are given in the preamble; (a) "***Cultural heritage*** *refers to monuments, groups of buildings and sites of heritage value, constituting the historic or built environment*"; (b) "***Recording*** *is the capture of information which describes the physical configuration, condition and use of monuments, groups of buildings and sites, at points in time, and it is an essential part of the conservation process*"; (c) "***Records*** *of monuments, groups of buildings and sites may include tangible as well as intangible evidence, and constitute a part of the documentation that can contribute to an understanding of the heritage and its related values.*". It is important to say that intangible aspects of cultural heritage are being stressed as an integral part of the documentation process.
- **The reasons for recording**
  This is the "Why" of the ICOMOS Principles. In this section it is explained why the recording of the cultural heritage is extremely important. Amongst the reasons mentioned are: knowledge acquisition, active encouragement and participation, informed management and control, and maintenance ensuring. Apart from this, the detailed recording is necessary for the appropriate information provision, in the suitable level of detail. Recording of cultural heritage is a priority.
- **Responsibility for recording**
  This is the explanation for the "Who". The second section of the ICOMOS Principles focalizes on the responsibility for recording cultural heritage and considers four factors: the necessary commitment, the adequate skills of the experts, the roles of the different specialties, and the leading role of the managers.
- **Planning for recording**
  Here is the "How", and planning as a stage is very important in the recording of cultural heritage. All the existing and archival sources of information should be found and inspected in detail to determine their adequacy state. Planning, as an essential part of recording, embraces also the optimal choice of the appropriate scope, as well as the level and methods of recording.
- **Content of records**
  When there are no records, it is like no recording process has been performed. This is the "What" in the ICOMOS Principles document. Any record should be identified on the basis of specific features. The location and extent of a cultural heritage object is an imperative need while creating records, and this is feasible by

using various means. New records should indicate the sources of all information while records should finally include specific information, which is mentioned in the document. The different reasons for recording reflect that different levels of detail will be required.

- **Management, dissemination and sharing of records**
  The last section of the ICOMOS Principles concerns the management practices suitable for protecting the records. The use of the appropriate technology is also considered. Accessibility and quality in terms of value, is of high importance; there is a need for sharing recording results with current and future researchers. The security and the accessibility of records are discussed in this last section.

2.3.6 Other ICOMOS Principles

There are also other ICOMOS Principles; it is interesting to draw attention with respect to documentation aspects, for example those ICOMOS Principles related to the different material elements.

- **Principles for the Preservation of Historic Timber Structures (1999)**
  The Principles for the Preservation of Historic Timber Structures were adopted by the ICOMOS 12th General Assembly in Mexico, in October 1999. Amongst the recommendations there is a section titled "Inspection, Recording and Documentation" that mentions: "Newby (2012). *The condition of the structure and its components should be carefully recorded before any intervention, as well as all materials used in treatments, in accordance with Article 16 of the Venice Charter and the ICOMOS Principles for the Recording of Monuments, Groups of Buildings and Sites. All pertinent documentation, including characteristic samples of redundant materials or members removed from the structure, and information about relevant traditional skills and technologies, should be collected, catalogued, securely stored and made accessible as appropriate. The documentation should also include the specific reasons given for choice of materials and methods in the preservation work.* OECD (2001). *A thorough and accurate diagnosis of the condition and the causes of decay and structural failure of the timber structure should precede any intervention. The diagnosis should be based on documentary evidence, physical inspection and analysis, and, if necessary, measurements of physical conditions and non-destructive testing methods. This should not prevent necessary minor interventions and emergency measures.*"

- **Principles for the Preservation and Conservation-Restoration of Wall Paintings (2003)**
  This document has been ratified by the ICOMOS 14th General Assembly, in Victoria Falls, Zimbabwe, in October 2003. It can be applied to wall paintings and Article 3 refers specifically to the need for documentation ... "*In agreement with the Venice Charter, the conservation-restoration of wall paintings must be accom-*

panied by a precise program of documentation in the form of an analytical and critical report, illustrated with drawings, copies, photographs, mapping, etc. The condition of the paintings, the technical and formal features pertaining to the process of the creation and the history of the object must be recorded. Furthermore, every stage of the conservation-restoration, materials and methodology used should be documented. This report should be placed in the archives of a public institution and made available to the interested public. Copies of such documentation should also be kept in situ, or in the possession of those responsible for the monument. It is also recommended that the results of the work should be published. This documentation should consider definable units of area in terms of such investigations, diagnosis and treatment. Traditional methods of written and graphic documentation can be supplemented by digital methods. However, regardless of the technique, the permanence of the records and the future availability of the documentation is of utmost importance."

- **Principles for the Analysis, Conservation and Structural Restoration of Architectural Heritage**
  It had been also ratified by the ICOMOS 14th General Assembly, in Victoria Falls, Zimbabwe, in October 2003. The document clearly specifies that "*all the activities of checking and monitoring should be documented and kept as part of the history of the structure.*"

2.4 Other International Initiatives

2.4.1 Docomomo International

Docomomo International (Docomomo 1988), or simply Docomomo, or International Committee for Documentation and Conservation of Buildings, Sites and Neighbourhoods of the Modern Movement, as it is the full title, is a non-profit international organization initiated in 1988 by Hubert-Jan Henket, an architect and professor, and Wessel de Jonge, an architect and research fellow, at the School of Architecture at the Technical University in Eindhoven, The Netherlands.

The organization is managed by a secretariat. In 2002, the Docomomo International secretariat relocated to Paris. In 2010, the Docomomo International secretariat relocated to Barcelona, and currently, Docomomo International is hosted in Lisbon. Docomomo holds biennial international conferences where people working in the documentation and conservation issues gather to discuss and exchange experiences, information and studies. The first conference was organized in 1990—Eindhoven, The Netherlands. In addition to that, the International Scientific Committee on Technology (ISC/T) organizes seminars covering various themes, such as restoration of reinforced concrete structures, wood and the modern movement, stone in modern buildings, etc.

According to the Docomomo Constitution (Docomomo 2010), the Committee has three general aims:

1. "*The exchange of know how and ideas in the field of Modern Movement architecture and design and its documentation and conservation.*
2. *To act as watchdog when examples of Modern Movement architecture and urban design are in jeopardy.*
3. *To stimulate the interest of the public in general and the proper authorities in particular in Modern Movement architecture and modern design; to make an international register of important Modern Movement buildings to be preserved and/or documented; to formulate new ideas for the future of the build environment based on past experiences of the Modern Movement.*"

2.4.2 RecorDIM Initiative

Between the years 1995 and 1999, a series of workshops organized by CIPA has recognized critical gaps and disharmonies in the field of cultural heritage recording, documentation and information management. This was observable between the interests and activities of 'Information Users' such as those of architects, archaeologists, heritage site managers, etc. and those of 'Information Providers', for example, photogrammetrists, surveyors, etc. In response to that situation, ICOMOS, CIPA and the Getty Conservation Institute (GCI) together created a partnership called the RecorDIM Initiative.

The RecorDIM Initiative vision was presented during the 18th CIPA International Symposium in Potsdam, Germany (18–21 September 2001). The following figure (Fig. 2.5) illustrates the RecorDIM Initiative vision and challenges as presented at that time, which consisted of 'bridging the gaps' between the information users and the information providers, by employing knowledge sharing, skills transferring as well as by integrating activities to raise the global level of conservation practice.

This group of international experts investigated several ways to strengthen the documentation building block of built heritage conservation through the development of tools and training and also through better communication between the 'Information Users' and the 'Information Providers'. The group worked and acknowledged a series of matters, including the need for a publication on principles and guidelines for recording and documentation of cultural heritage. The outcome of this attempt was the book titled: Recording, Documentation, and Information Management for the Conservation of Heritage Places: Guiding Principles (Letellier et al. 2007).

This book provides a complete overview of the fundamental principles and guidelines for cultural heritage monuments and sites documentation. With this book, the authors aspire to provide help to the cultural heritage stakeholders, managers and decision makers in understanding their roles and their duties to deal with this extremely important activity.

2.4.3 The London Charter

The London Charter (2009) is tackling issues related to the computer-based visualization of cultural heritage. It was born in an international scientific context seeking to establish the necessary requirements to verify that a *3D* visualization of cultural heritage is credible. The question of transparency of the various *3D* visualization applications for cultural heritage is of high importance as a scientific discipline (Beacham et al. 2006; Hermon et al. 2007).

The main objective of the London Charter was to invert the authority principle in the generation of virtual models. According to this, depending on the person who created the model, it delectated more or less research-based status. The authority principle has been substituted by the scientific and research-based method, according to which the virtual models should be described by a set of data and information, such as metadata and paradata, in order to make easier their validity and evaluation by independent experts.

The London Charter was not undertaken to set in motion new proposals but rather to bring together the main inception already published by several authors; an inception not fully absorbed by a broad part of the scientific community across the world. This is an explanation why the "Charter" setting was used, instead of publishing a new article. It was assumed the most appropriate tool to assure its spreading more widely and discussion in the international society of experts who practice in 3D representations and visualizations.

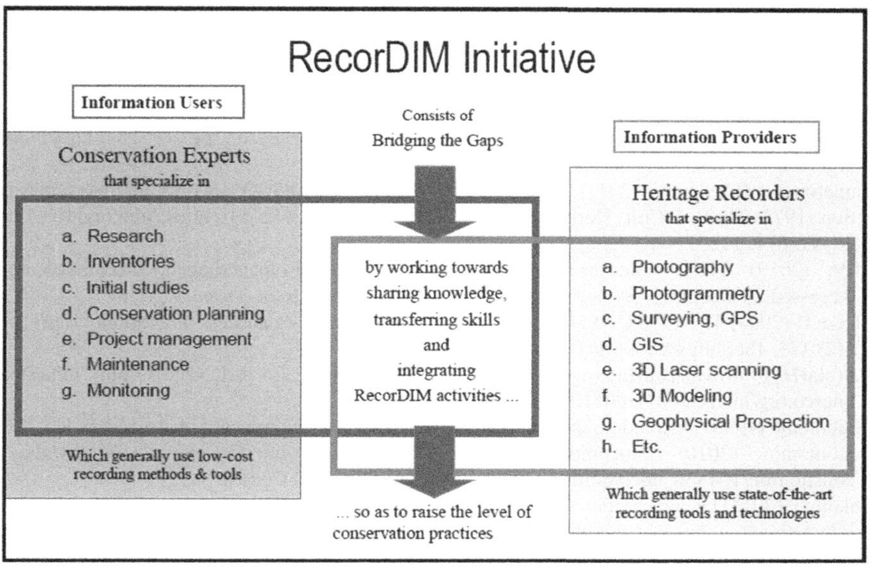

Fig. 2.5 The RecorDIM Initiative vision and challenges (Letellier 2002)

In spite of the fact that the term "Charter" is officially used for those documents approved by ICOMOS General Assembly, the authors of the London Charter decided to use the same term pursuing the importance of this initiative. The London Charter has not been ratified by the ICOMOS General Assembly.

2.4.4 The Seville Principles

The International Principles of Virtual Archaeology, widely known as the Seville Principles (2011), is a set of particularities of the London Charter. This document took its name from the city of Seville in Spain, where it had been initiated (Bendicho and Lepez-Menchero 2013).

The London Charter is a set of recommendations which are relevant to the cultural heritage in general, while the Seville Principles focused on archaeological heritage as a clearly defined component of cultural heritage. Following the devising of ICOMOS, the Seville Principles were set in the category of "principles", a level below "charter". However, the Seville Principles still have the fate of the London Charter as they have not been approved by the ICOMOS General Assembly.

References

Beacham R, Denard H, Niccolucci F (2006) An introduction to the London charter. In: The e-volution of information communication technology in cultural heritage - joint event CIPA/VAST/EG/EuroMed event, Archaeolingua, Budapest

Bendicho V, Lepez-Menchero M (2013) International guidelines for virtual archaeology: the seville principles. In: Corsi C, Slapšak B, Vermeulen F (eds) Good practice in archaeological diagnostics: non-invasive survey of complex archaeological sites. Springer, Berlin, pp 269–284. https://doi.org/10.1007/978-3-319-01784-6_16

Cameron C, Rössler M (2011) Voices of the pioneers: UNESCO's world heritage convention 1972–2000. J Cult Herit Manag Sustain Devel 1(1):42–54. https://doi.org/10.1108/20441261111129924

CIPA (2019) CIPA - heritage documentation. http://www.cipaheritagedocumentation.org/. Accessed 23 Dec 2019

Cleere H (2004) The 1972 UNESCO world heritage convention - a auccess or a failure? Herit Soc 4(2):173–186. https://doi.org/10.1179/hso.2011.4.2.173

Daifuku H (1998) Museum international: fiftieth anniversary issue. Ed. by UNESCO. http://unesdoc.unesco.org/images/0011/001105/110513e.pdf#xml

Docomomo (1988) Docomomo international. https://www.docomomo.com. Accessed 28 Jan 2018

Docomomo (2010) Docomomo constitution. https://www.docomomo.com/pdfs/about/constitution/104836_docomomoconstitution.pdf. Accessed 29 Dec 2018

Hassani F (2015) Documentation of cultural heritage techniques, potentials and constraints. In: The international archives of the photogrammetry, remote sensing and spatial information sciences XL-5/W7, pp 207–214

Hermon S, Sugimoto G, Mara H (2007) The London charter and its applicability. In: Arnold D, Niccolucci F, Chalmers A (eds) The 8th international symposium on virtual reality, archaeology and cultural heritage VAST (2007), Brighton, pp 11–14

Iamandi C (1997) The Charters of Athens of 1931 and 1933: coincidence, controversy and convergence. Conserv Manag Archaeol Sites 1(2):17–28. https://doi.org/10.1179/135050397793138934

ICOMOS, Australia, (2013) The Burra charter, Australia. http://australia.icomos.org/publications/burra-charter-practice-notes. Accessed 03 Feb 2018

ICOMOS (2018) ICOMOS - international council on monuments and sites, France. https://www.icomos.org/en. Accessed 05 Feb 2018

ISPRS (2018) ISPRS - the international society for photogrammetry and remote sensing. http://www.isprs.org/. Accessed 05 Feb 2018

Jokilehto J (1998) "The context of the Venice Charter (1964). Conserv Manag Archaeol Sites 2:229–233

Letellier R (2002) ICOMOS - CIPA - GCI: 5-year RecorDIM initiative. Preliminary plan 2002–2003

Letellier R, Schmid W, LeBlanc F (2007) Recording, documentation, and information management for the conservation of heritage places. Guiding principles. The Getty Conservation Institute. 151 pp. ISBN: 978-0-89236-925-6

Locke WJ (1904) Recommendations of the Madrid conference. The 6th International congress of architects. In: The architectural journal: being the journal of the royal institute of british architects (RIBA) Third Series XI, pp 343–346

LondonCharter (2009) The London Charter: for the computer-based visualization of cultural heritage. Version 2.1. http://www.londoncharter.org/. Accessed 30 Dec 2018

Newby PRT (2012) Photogrammetric terminology. Photogramm Rec 27(139):360–386. https://doi.org/10.1111/j.1477-9730.2012.00693.x

NHPA (1966) National historic preservation act - public law, pp 89–665. http://www.achp.gov/. Accessed 09 Feb 2018

NPSc (2017) The secretary of the interior's standards for the treatment of historic properties with guidelines for preserving, rehabilitating, restoring and reconstructing historic buildings

OECD (2001) Glossary of statistical terms. https://stats.oecd.org/glossary/detail.asp?ID=1235. Accessed 02 Mar 2018

OSI (2007) The open source definition. https://opensource.org/osd. Accessed 02 Mar 2018

Patias P (2004) 35 years of CIPA. In: ISPRS archives XXXV.B5, pp 834–838. http://www.isprs.org/proceedings/XXXV/congress/comm5/papers/665.pdf

Rodwell D (2012) The UNESCO world heritage convention, 1972–2012: reflections and directions. Hist Environ: Pol Pract 3(1):64–85. https://doi.org/10.1179/1756750512Z.0000000004

UNESCO (1972) The world heritage convention. http://whc.unesco.org/en/convention/. Accessed 06 Mar 2018

UNESCO (2017) The operational guidelines for the implementation of the world heritage convention. http://whc.unesco.org/en/guidelines/. Accessed 06 Mar 2018

Vigneron S (2016) From local to world heritage: a comparative analysis. Hist Environ: Pol Pract 7(2–3):115–132. https://doi.org/10.1080/17567505.2016.1172779

Waldhäusl P (2004) Hans foramitti a pioneer of architectural photogrammetry (1923–1982). In: ISPRS archives XXXV.B5, pp 828–833. http://www.isprs.org/proceedings/XXXV/congress/comm5/papers/664.pdf. Accessed 20 Feb 2018

Chapter 3
The Need for Documentation

Abstract This chapter undertakes to respond to the need for cultural heritage documentation. Foreseeable or indeterminate situations, like natural and human-made threats, are becoming serious obstacles of preserving all cultural heritage wealth, regardless the type. What is documentation? This is the first question answered in this chapter. The five "W"s (Who, What, Where, When, Why) and the one "H" (How) could be considered as a "documentation narrative" to define a workflow. Data processes are important building blocks of the documentation process. Data exist in many different places and in many different types. The open data framework is also presented in this chapter, as well as issues related to the Intellectual Property Rights on cultural heritage. The regulatory framework and the international treaties are also addressed. A general overview of the sensors and systems used for the documentation of cultural heritage, in terms of active and passive sensors, and other systems are presented. Tools and software, both commercial and open source, are illustrated in this chapter. The chapter concludes with two different examples based on human threats and natural disasters.

Although we ought to preserve all cultural heritage wealth, regardless the type, like a monument, an archaeological site or a cultural landscape, moveable or immovable, it seems to be impossible to reach this goal. Predictable or uncertain conditions, like natural and human threats (Sect. 3.3), are becoming serious obstacles of this endeavor. It looks like we are unable to save everything.

Documentation as an integral part of preservation cycle should be our choice. Cultural heritage should be documented before any damage or loss. A complete record of any object, monument or site, containing of all the proper data and information, is not only able to support any virtual or physical reconstruction process if decided, but also it is capable for knowledge transfer to the generations coming. Documentation is an irreplaceable link to the entire cultural heritage restoration, conservation or preservation process. Certainly, it can help preserving cultural heritage from being devastated or from falling into lethe, and it functions towards communicate and raise

E. Stylianidis, *Photogrammetric Survey for the Recording and Documentation of Historic Buildings*, Springer Tracts in Civil Engineering,
https://doi.org/10.1007/978-3-030-47310-5_3

Fig. 3.1 The destroyed minaret of the Umayyad mosque of Aleppo. (*Source* Gabriele Fangi, Wissam Wahbeh/CC3 Creative Commons)

awareness, directed not only to conservation experts and professionals, but also to the wider public. Such a work is able to transfer the real message for the value, character, and significance of cultural heritage.

In Syria (Fig. 3.1), Iraq, Afghanistan, and so many other places around the world, cultural heritage has fallen prey to destruction from war conflict and terrorist attacks, by heavy artillery, explosive attacks and other violent and illegal actions. However, there is another important viewpoint for civilized humanity. Although the antiquities devastation may appear insignificant compared with current carnages, the irreparable impairment to cultural heritage might have a severe impact on people who survive a war. Especially in the cultural identity of people suffering the consequences of a war conflict. The destruction of cultural heritage is not just the damage of exceptional monuments, sites or objects, but rather a severe disruption of a vigorous past with its existing historical places, customs and behaviors. Saving cultural heritage and helping people who are struggling to live through a war is an obligation of the civilized world.

3.1 Defining Documentation

What is documentation? Broadly speaking, it can be said that documentation is a process which comprises four main undertakings (PAPS—Plan, Acquire, Process, Store):

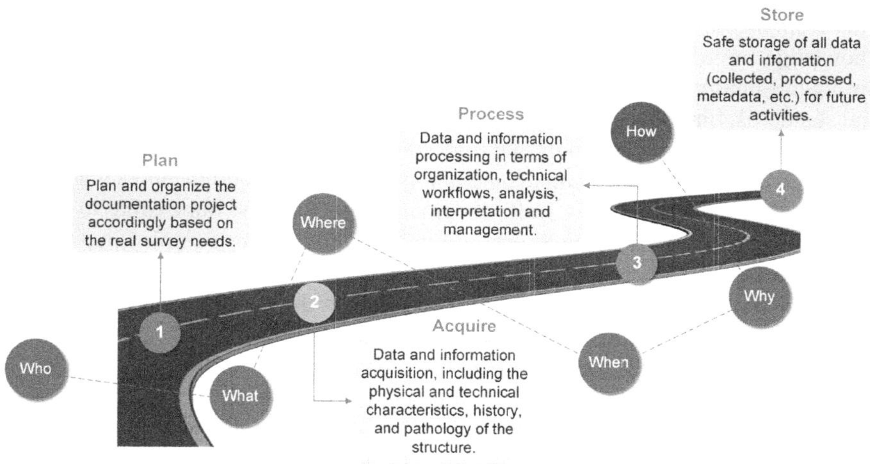

Fig. 3.2 Documentation pipeline, embraced by the 5W1H workflow

1. **Plan**: planning and organization of the scheduled tasks to gather information and data with respect to the cultural heritage object under investigation.
2. **Acquire**: data and information acquisition, including the physical, technical and descriptive characteristics, history, and pathology.
3. **Process**: data and information processing in terms of organization, analysis, interpretation and management.
4. **Store**: safe storage of all the collected and processed data and information, regardless of the scope of the future activities.

The respective figure (Fig. 3.2) illustrates the documentation process as a sequence of the four consecutive steps (PAPS) that are taking place as modules of an integrated pipeline. In addition to this four-stage pipeline, the whole process is connected with the 5W1H workflow (Sect. 3.1.1), i.e. the five "W"s (Who, What, Where, When, Why) and the one "H" (How).

There is no need to convince anyone of the importance of *planning*. Planning should be an imperative preparatory phase for every planned activity. The introduction of the term "documentation planning", redefines it and explains why it is important for all professionals working in the domain of restoration/conservation/preservation of cultural heritage, to invest their time planning for the real needs and their daily business. Planning is one of the most thoughtful in character project management and time management techniques of carrying out a particular task or a package of tasks. Planning is preparing a suite of action steps to succeed in the set goal which is the documentation of a specific cultural heritage object or structure. Effective planning implementation is able to reduce the necessary time and effort of achieving the desired result. In addition to that, planning cultivates team building and a spirit of cooperation within the interdisciplinary documentation team. When the plan is accomplished and shared among the team members, everyone knows what their responsibilities are, and how other team members need their support, skills and knowledge in order to finish the assigned tasks. Everything should be considered during this phase: human and funding resources, equipment, technical specifications and standards, limitations and restrictions in terms of time, legal and ethical aspects.

It is obvious that without data and information there is no ground for documentation, even as a single term. *Data acquisition* is incredibly important and a mandatory process, because without this stuff, documentation is almost impossible to further proceed. Datasets are so large in size (digital) and sometimes complex due to their heterogeneous character; for example, compare the descriptive historic data and *3D* laser scanning data for a historic building. Whatever the cultural heritage object is, small or huge, certainly, data is the critical parameter in changing the way of approach as well as the quality of the documentation outcomes. It is easy for anyone to get caught up in simply saying "collect as much data as you can". However, data management begins by being aware and comprehending what inquiries and outcomes are requested in a documentation campaign. Connecting the documentation mission with outcomes data and single metrics can be tough, but it is a necessary process to transform the data management system into a powerful tool for organizational and scheduling performance. Data only matters if it translates into action towards real delivery of products. Data only matters if it translates into real documentation archives, ready for further use.

Data is not always in processed form, and thus, data processing is necessary to compile data in a desirable format or outcome. *Data processing* is simply the conversion of measurements, raw data or any other data source to meaningful information through a series of necessary actions, i.e. it is a process. Data is handled properly to generate outcomes for the documentation, while analyzing data that has not been closely screened for any type of problem can derive highly misleading outcomes that to a great degree are contingent on the data quality. Processing is when the data is exposed to miscellaneous means and methods of handling. Many software programs are able to be used for processing large data volumes into very short time. Data processing leads to outcomes and products, then the processed information is transmitted to the users. Outcomes and products are presented in various types and formats, from reports to representative *3D* models. Analysis and interpretation are taking place at this stage. Outputs need to be interpreted so that they can make available for use meaningful information that will guide future decisions.

When it comes to *data storage*, there is no 'one-size-fits-all' solution, especially in the case where both analogue and digital data and information exist. Before the decision where and how to store this extraordinary amount of priceless items, analogue and digital information, structured and unstructured data, the documentation team needs to savvy the quantity and data type available, along with the intrinsic and extrinsic motivation behind storing data and information. Even though all data is not created equal, the data value is critical for defining the restoration and conservation strategy for the historic object under study. Setting the justified data retention policies is an indispensable action for internal data governance, legal compliance, use by other teams or the wider public, but more importantly for future use. In the cultural heritage preservation domain, all data and information must be retained for ever.

Make sure that all data and information is secure. When managing such important and historic data, security has to be a high-level priority. In this sense, security comes two-fold: data and information have to be secure both physically and virtually. It is a must backing up all digital data and information in a secure location outside the premises. In the event of an unexpected or natural disaster, everything should be recreatable.

3.1.1 The 5W1H of Documentation

The five "W"s (Who, What, Where, When, Why) and the one "H" (How) (Fig. 3.3) could be assumed as a "documentation narrative". In fact, they are questions whose replies are deemed fundamental in data and information collection or problem solving, like in the case of cultural heritage documentation. They constitute a prescription for getting the thorough account of past and current tasks on the subject. According to the 5W1H principle, a report as an outcome of this process can only be considered complete if it respond to all the questions starting with an interrogative word, for the five "W"s and the one "H".

With the initiation of a new documentation project, the responsible documentation team should start thinking about the appropriate decisions. Will it be enough to

Fig. 3.3 The 5W1H of documentation

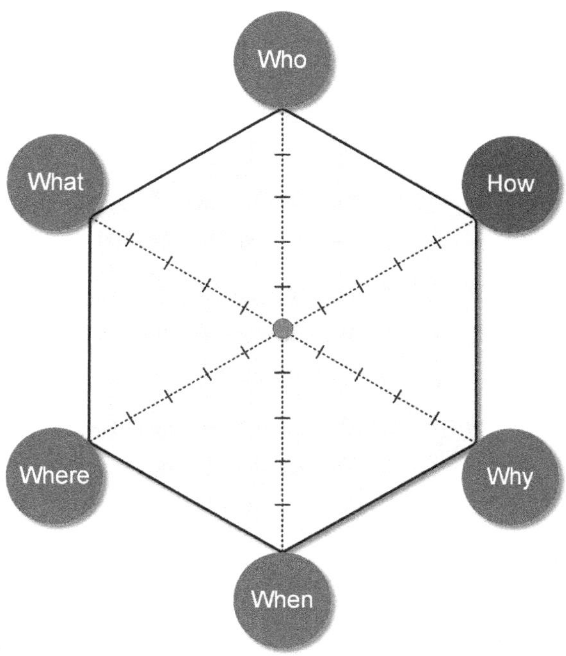

address only few of the "W"s? And what about "H"? One of the reasons why many documentation campaigns fail to fulfill all the requirements of success is that the team members and the team leader assume they have answered all the questions, when the reality is that none of them has been duly tackled.

Make sure you have asked yourself and all team members all the questions, for all the "W"s and the "H" too, on purpose and carefully, before even considering that anything is simple or obvious. For instance, respond to queries like the following:

- **What** are the issues this historic building is facing? Have you thought about them cautiously enough to place them into some order of priority, supposing that tackling them all at once is likely to be more than you can manage? Are you clear about exactly what these issues are? Do you perceive the intended meaning of all as fully as possible?
- **Why** does it make sense that you should deal with these issues? Is there something you wish to reach a desired objective; or something you reckon you need to change?
- **When** should you start the project? Is now the right time? Is it something urgent? Are the conditions favorable enough? Are you in risk of haste into short-term actions when a long-term approach is required?
- **Where** should you start this documentation project? Which particular part of the problem should be tackled before all others? Is it the most important or the most pressing?

Table 3.1 An example 5W1H scenario for a historic building documentation

Indicative 5W1H questions for a historic building documentation	
Who	Who is preparing the drawings?
	Who is leading the image acquisition task?
	Who is responsible for the archival search?
What	What kind of building is this?
	What filetypes should we use for the images?
	What is the optimum timeframe for the measurements every day?
Where	Where is the historic building?
	Where are the archives with the old photographs?
	Where shall we store the topographic measurements?
When	When is the deadline for project delivery?
	When shall we complete the laser scanning task?
	When do we have to finish the historic maps scanning?
Why	Why should we document this historic building?
	Why was a tight project completion program chosen?
	Why do we measure this part of the historic building?
How	How will the documentation be implemented?
	How else might it be done?
	How do we know if this information is correct?

- **Who** do you need on the documentation team? Professionals, technicians, workers? Have you specified the roles for each one of the team members? Is everyone aware of their responsibilities and duties?
- **How** will you organize and perform the documentation of the historic building? How will you choose the best approach? Are you aware of all the skills, techniques and technologies will you need? Have you learned from others' experiences to facilitate your work?

To better understand the function of the 5W1H workflow, the following example scenario for the documentation of a historic building is provided. Three indicative questions are given for each one of the 5 "W"s and the 1 "H" to stress theoretical and practical needs for such a specific project (Table 3.1).

3.2 The Need for Documentation

Frequently, as a logical consequence, cultural heritage brings to mind artifacts, i.e. real objects such as paintings or sculptures, monuments, historic buildings, as well as heritage landscapes and archaeological sites. However, cultural heritage is even wider than this concept. Gradually, cultural heritage imbibes, and new elements are considered as part of this ecumenic value, like historic maps, old documents or traditional surveying instruments. Nowadays, even more, where cultural heritage is in danger, old towns, shipwrecks and the natural environment formally belong to the cultural heritage family.

Cultural heritage does not only refer to material objects that we can perceive with the eyes and touch with the hands. It is also made up of intangible elements and attributes, such as traditional craftsmanship, music and dancing, rituals, knowledge and skills transmitted from artisans to the next generations.

Whatever it is, cultural heritage is the mirror of world history. The precise and sober study of this legacy is a commitment of the civilized world to mankind's past and future. Esteem to cultural heritage is dated back for centuries now, but since then, this is not always the case, considering the destruction of monuments and sites during wars or terrorist attacks.

However, cultural heritage is not just a collection of cultural objects, attributes or traditions that was delivered from the past and the previous generations. It is also the outcome of the action of the careful selection between memory and oblivion. These are characteristics and attitudes of human society that are based on anthropological, sociological, cultural, and political context. People and human societies decide what to bequeath to the next generations.

The protection and preservation of cultural heritage is not a new problem for the civilized world. The countries and the peoples are those creating culture, while at the same time, have the responsibility to respect and safeguard every single part of cultural heritage. They have the tools and the means, the laws, treaties and conventions, in national and international level.

In this context, the documentation of cultural heritage is an integral part of the restoration, protection, conservation and preservation process. As already explained, (Sect. 3.1), documentation includes the planning, acquisition, processing and storage of all possible data concerning the cultural heritage under study that may help towards its safeguarding, now or in the future. Historic, architectural, structural, administrative data and information, old photographs, images, drawings, sketches, measurements, etc., which can be collected or found in different sources, can contribute to the documentation of the cultural heritage object.

The documentation of cultural heritage is very important for many reasons, where the most important are:

- By documenting cultural heritage, we are assessing the values and significance of the heritage itself across the different epochs.
- Most of the time, documentation is taking place as a forerunner of conservation and a necessary action to guide and backing up this process.
- It is a supporting tool for monitoring and management.
- Contributes to the creation of a valuable record for archiving purposes.
- Communicating the importance of the cultural heritage object under study to the special and wider public.

The international framework for the documentation of cultural heritage has already been discussed in the previous chapter. The need for documentation infused the vast majority of the ICOMOS Charters and Principles. An indicative snapshot from three international references to the need for documentation is provided in Fig. 3.4.

3.2.1 Data and Processes

In the context of historic building preservation, documentation typically refers to a complete record, or evidence that the building has an objective reality. Record means data and the spectrum of data (Fig. 3.5) that is captured differs depending the specific scope of the project. Therefore, the documentation components consist of an extremely important source in understanding the building, its characteristics and the definition of actions to be taken for its preservation. In fact, it is the observable evidence that the conservators stand in to validate motive, interpretation and finally the actions to take place in the building. Physical data may also serve as the one and only piece of information that indicates the historic building's existence and condition at a specific time. The importance of the data can determine the unique real facts about the existence of the building. In addition to that, it may also serve as a record of actual experiences related to a specific period in history, where the practices and documentation techniques become equally as important as the real data and content.

Usually, the built environment, and thus, the inherent historic buildings, is recorded through old photographs, images, (laser scanning) point clouds, drawings,

CONSEIL INTERNATIONAL DES MONUMENTS ET DES SITES — ICOMOS — INTERNATIONAL COUNCIL ON MONUMENTS AND SITES

INTERNATIONAL CHARTER FOR THE CONSERVATION AND RESTORATION OF MONUMENTS AND SITES (THE VENICE CHARTER 1964)

Article 16.

In all works of preservation, restoration or excavation, there should always be precise documentation in the form of analytical and critical reports, illustrated with drawings and photographs. Every stage of the work of clearing, consolidation, rearrangement and integration, as well as technical and formal features identified during the course of the work, should be included. This record should be placed in the archives of a public institution and made available to research workers. It is recommended that the report should be published.

UNESCO

UNITED NATIONS EDUCATIONAL, SCIENTIFIC AND CULTURAL ORGANISATION

CONVENTION CONCERNING THE PROTECTION OF THE WORLD CULTURAL AND NATURAL HERITAGE

Adopted by the General Conference at its seventeenth session Paris, 16 november 1972

Article 11

1. Every State Party to this Convention shall, in so far as possible, submit to the World Heritage Committee an inventory of property forming part of the cultural and natural heritage, situated in its territory and suitable for inclusion in the list provided for in paragraph 2 of this Article. This inventory, which shall not be considered exhaustive, shall include documentation about the location of the property in question and its significance.

CONSEIL INTERNATIONAL DES MONUMENTS ET DES SITES — ICOMOS — INTERNATIONAL COUNCIL ON MONUMENTS AND SITES

CHARTER FOR THE CONSERVATION OF HISTORIC TOWNS AND URBAN AREAS (WASHINGTON CHARTER 1987)

METHODS AND INSTRUMENTS

5. Planning for the conservation of historic towns and urban areas should be preceded by multidisciplinary studies.

Conservation plans must address all relevant factors including archaeology, history, architecture, techniques, sociology and economics.

The principal objectives of the conservation plan should be clearly stated as should the legal, administrative and financial measures necessary to attain them.

The conservation plan should aim at ensuring a harmonious relationship between the historic urban areas and the town as a whole.

The conservation plan should determine which buildings must be preserved, which should be preserved under certain circumstances and which, under quite exceptional circumstances, might be expendable.

Before any intervention, existing conditions in the area should be thoroughly documented.

The conservation plan should be supported by the residents of the historic area.

Fig. 3.4 Three indicative examples from international references to the need for documentation; two ICOMOS Charters (1964, 1987) and one UNESCO Convention (1972). (*Source* ICOMOS)

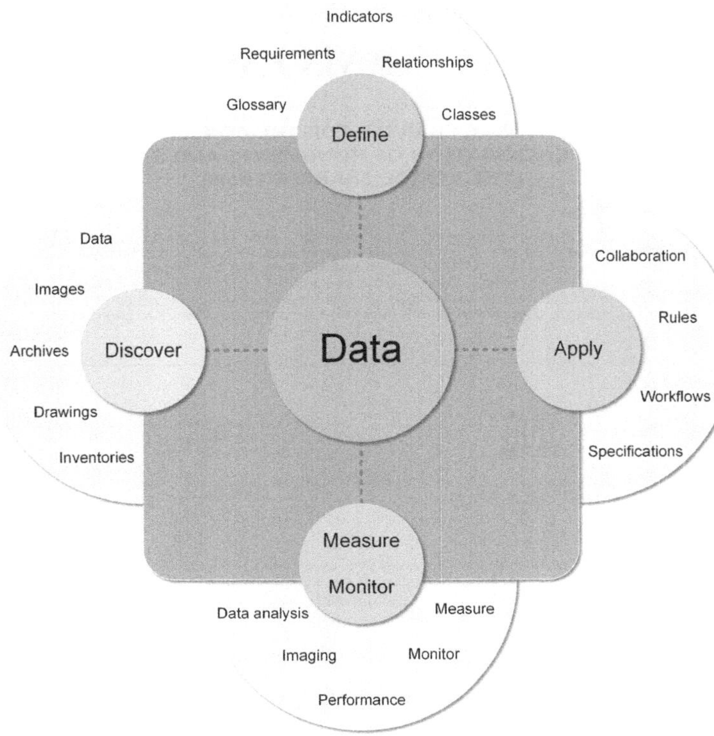

Fig. 3.5 Data and processes

reports, etc. Each one of these respectively reflects to the methods, techniques and technologies used, that has both advantages and disadvantages. Graphic outcomes such as images and their derivatives (e.g. an orthoimage), sketches, drawings, etc., are extremely useful graphical and schematic representations of the historic building. But also, the written documents, such as the technical reports, are very useful to communicate the information related to the building, its characteristics and the findings. A report and the wording used is not capable of adequately describing the recording and documentation of a historic building. A measured drawing has the ability to quickly and accurately record and transmit the physical connection of building components and their construction. Graphic products are also needed to provide all the required information and the building condition. A written report is a more appropriate choice for presenting the study in a narrative and descriptive context. Both should coexist harmoniously and tightly in the interest of historic recording and documentation. This integrated documentation should be done aright, so its usefulness has the appropriate historicalness and sustainability for the building and the next generations.

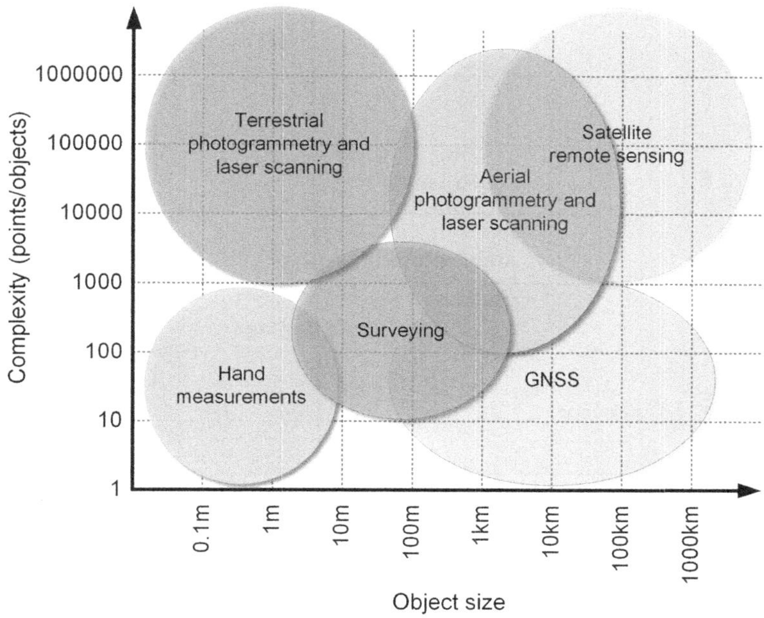

Fig. 3.6 2D/3D surveying techniques defined by object complexity and size

There are many and different available *2D* and *3D* surveying methods and techniques contributing to the recording and documentation of cultural heritage. It is obvious that in the easy cases, the project team can use simple tools, equipment and surveying methods to avoid high equipment costs. However, in more complex cases, which seem to be generally the regular case, several other methods should be used. These methods are accompanied with the appropriate equipment and workflows, too. This approach of the various surveying methods depicted by Boehler and Heinz (1999), where the result of Fig. 3.6 is inspired; they used a more simplified schematic representation. They also endorsed that the suitable surveying methods to be used can be found by considering the object size and the complexity of the object. On the one hand, the object size is highly related to the scale and the accuracy of the products; on the other hand, the complexity of the survey can be expressed by the number of recorded points.

They recommend (Boehler and Heinz 1999) that besides the size and the complexity of the object under survey, a project team should think carefully of other factors that may have the capacity to have an effect on the most favorable method to be chosen:

- The required accuracy for the project needs.
- The selected method or part of it may be forbidden (e.g., no permission to fly with an Unmanned Aerial Vehicle (UAV) to take aerial images).

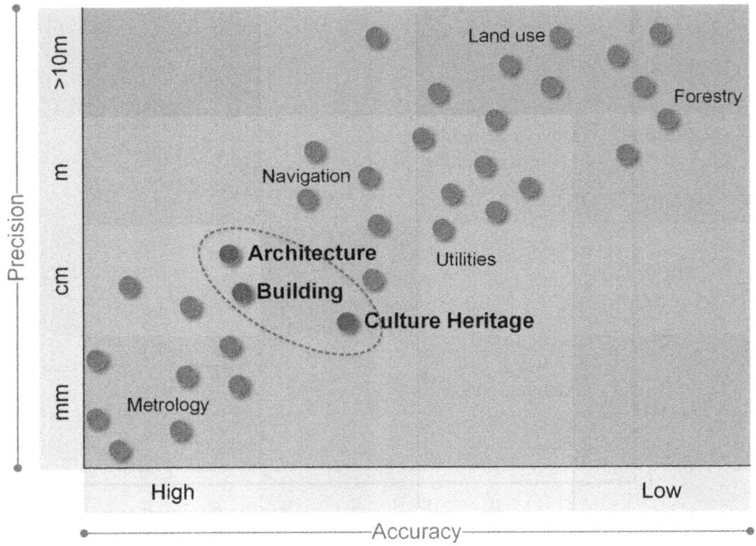

Fig. 3.7 Precision versus Accuracy and cultural heritage

- The availability of the instruments required to be used and the occurrence of a power supply.
- The accessibility of the object (e.g. a building with facades that are not equally accessible for measurements).
- The availability of preferably located vibration-free observation and measurement stations.
- The permission to touch the object which is under recording and documentation.

Nowadays, the industry of ICT, sensors, imaging, and geospatial technologies is moving from position to precision, and architecture, building information and cultural heritage are placed in a specific area, as illustrated in Fig. 3.7. In a wider view, one can endorse that this group of heritage-related disciplines can be put in a particular position very close to high (precision and accuracy). It is obvious that in this super technological ecosystem with so many cross-cutting technologies, the most significant transitions in the geospatial and measurement domain are almost impossible to originate from one and single technology. The combination and involvement of several technologies is the game changer towards new and novel products. It is enough to look at only three different, and so characteristic and representative, products of the specific sector: a total station, a laser scanner and a UAV.

This digital community that embraces all ecosystems around geospatial technologies, if they can be put under one umbrella, is one of the best examples of research, technology, development and innovation (RTDI) principle. The ubiquitous nature of mapping and simultaneous imaging is only possible and powerful due to the booming in mobile technology, the miniaturization of the imaging sensors for different

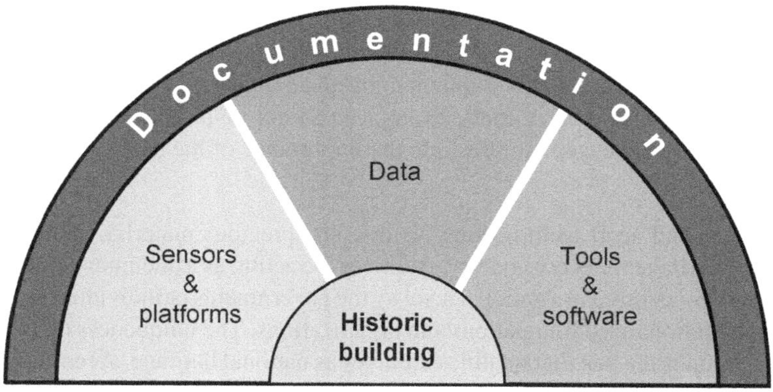

Fig. 3.8 The building blocks for the documentation of a historic building

types of cameras, the universal availability of global positioning (GNSS), the speed increase of the data and wireless telecommunication networks, and the ability to store and distribute large datasets over the cloud. The globe is hosted in our mobile devices.

As with all technologies, the state and the quality that they produce becomes higher as progress and technological innovations and improvements are coming to the foreground. This is the merit of the productivity that technology is generating. Look at the example of sensors which is so cogent. Sensors are not just processing the information of particular place, the properties and the images of this particular place, but also the time that the information was captured. This temporal dimension, which is so important for the documentation of cultural heritage, provides a value of great significance and a principal component for many applications and services.

Data is one of the most important building blocks (Fig. 3.8) towards the documentation of a historic building or any other cultural heritage object. In order to continue considering historic buildings along with the other historic environment as the tangible evidences of history, data is the uncontested and primary source. This reflects both to archival and modern data. The survey and recording are the tools for data collection, and thus, a prerequisite for any documentation process of cultural heritage. No further reliable action can be assumed without having an accurate and complete data collection.

3.2.1.1 Where Is the Data?

Apart from the data collected at the time of building recording and documentation, there is an incredible wealth of archival material that seems to be hidden in various places; known but also unknown. That is why it is necessary to have an excellent knowledge of the administration environment and the existence of potential data sources. But also smart and effective ways to search, acquire or even download material, information and data.

So, where is the data? What are the potentially optimal sources for searching and finding such information and data? The following list gives a picture of the reality that emerges from the experience of implementing and managing such projects. For each of the sources analyzed next, many examples can not be provided. The paradigms provided as indicative cases to illustrate the importance of the listed sources.

- **Archives**
  By nature, and apart from hosting historic and precious material, archives are a national heritage for every country. Archives are acting as a documentary evidence of activities occurred in a national level by the governments or individual persons or even institutionalized foundations and organizations. The uniqueness of the stored information is the one that qualifies archives as national heritage. According to the Society of American Archivists (SAA) (SAA 2018), archives are documents made or received and accumulated by a person or organization while carrying out affairs and preserved because of their persisting value. Frequently, archives consist of accumulation of documents and are managed as such, despite the fact that archival foundations often maintain discrete items that must also be treated systematically within the descriptive system of the foundation. Apart from that, archives can also be used as the term to make reference to an organization or a program liable for the selection, maintenance and their usage. One more expression can also refer to a repository, building or place responsible for their storage, preservation and usage. In an effort to recognize what makes different records and archives, (Williams 2006) argues that the generally accepted difference between them, at least for most anglophones, is that each term has a different meaning:
  - *Records* as a term has been applicable to the outputs of current and ongoing activity.
  - *Archives* as a term has been stated as referring to any records with long-term persisting value that have been kept either because they may be essential for continuing organizational purposes to their creating organization or since they have addable research value.
- **Institutions and Foundations**
  The world's cultural heritage is part of the global cultural diversity through the different epochs, regions and religions. It concerns also the various historic, artistic and social distinctive appearance of a culture. They make available for use opportunities to recognize the full worth of the different cultural traditions, but also demand preservation strategies so as to continue to exist, ahead the new coming epochs. Cultural heritage, as the process of making known humans' thoughts, feelings and existence, constantly inspires many people to offer more than others, not only because they need to offer more but also because they have more to offer. Therefore, responsibility for the protection of cultural heritage should be spread to as many as possible in order to maximize the impact worldwide and to build the bridge between the past, present and future.
  The universal significance of cultural heritage preservation led many passionate people to offer, to some extent or even the whole, their fortunes to this ultimate

goal. They contributed in motivating the creation of networks of people and other mechanisms in cultural heritage preservation domain, and to offer there where the preservation is essential for the future generations. And there are many such examples of persons having a catalytic effect all around the world, and that is not possible to be introduced within the framework of this book.

In the name of these important persons and donors, different foundations and institutions were established, especially by granting a very rich endowment. Many of these institutions have enormous collections and an incredible wealth of data and information, mostly in the possession of donors from previous years. All this huge number of objects, collections, books, or any other type of information was given to the institution. But it is not just these achievements. Due to their sound financial background, these foundations skillfully managed to acquire more data and more valuable objects. They carried out projects all around the world and collected very important information, while they have also received other donations from smaller donors. This has made these foundations important databanks and owners of cultural objects worldwide and thus, a valuable source of information for cultural heritage documentation and preservation projects.

The J. Paul Getty Trust is the world's largest cultural and philanthropic institution devoted to visual arts: to its presentation, conservation and interpretation. Getty has four building blocks, namely the Getty Conservation Institute, the Getty Foundation, the J. Paul Getty Museum, and the Getty Research Institute, which implement a collective and individual work in Los Angeles, US, which is its base, but also all around the world. It pursues to serve both the wider concerned public and professional communities in order to further the progress in an absolutely necessary civil society through an understanding of the visual arts. Getty has an extensive portfolio of activities covering conservation and training programs, publications and exhibitions, or offering grants, a real contribution in conservation practice and art history research (Getty 2018).

The Aga Khan Development Network (AKDN) was founded by the Highness the Aga Khan, the 49th hereditary Imam (Spiritual Leader) of the Shia Imami Ismaili Muslims. AKDN and its founder dedicated their efforts to improving the quality of life of the most vulnerable populations, giving special prominence to the view of Islam as a faith that teaches compassion and tolerance and that supports human dignity. Amongst the many sectors AKDN is targeting are architecture, culture, education and historic cities. For instance, the Aga Khan Trust for Culture (AKTC) has shown how culture can be a catalyst for improving the quality of life. Through its programme, AKTC is targeting the preservation and revitalization of cultural assets and cultural traditions, the creation of education programmes that bring up mutual understanding, and the reconnaissance of architectural excellence that in a positive way impacts the manner in which people live, work and interact (AKDN 2019).

The J. M. Kaplan Fund is a philanthropic organization over three generations. The Fund was established in 1945 by Jacob M. Kaplan, and it is funding programs in three main areas, namely social justice, environment and heritage conservation. The Fund is committed to preserving and conserving cultural heritage as it has long believed in the value of cultural heritage. The projects supported by J. M. Kaplan Fund are either in the US or in the Mediterranean basin, where they helped conserve important cultural assets. The Heritage Conservation program of the Fund is currently focused on areas such as the conservation of sites of Greco-Roman antiquity, the protection of cultural heritage sites threatened by armed conflict, and the preservation of sites that can elevate and inform heritage practice in the US (JMK 2019).

- **Crowdsourcing**
  The term coined jointly by Jeff Howe and Mark Robinson, editors at Wired magazine, in 2006 (Ridge 2014). Originally, the two editors described how the enterprises were utilizing the Internet to "outsource work to the crowd", which rapidly led to the term "crowdsourcing". Howe (2006a) published a definition for this term in an article under the title, "Crowdsourcing: A Definition"; "*Simply defined, crowdsourcing represents the act of a company or institution taking a function once performed by employees and outsourcing it to an undefined (and generally large) network of people in the form of an open call. This can take the form of peer-production (when the job is performed collaboratively), but is also often undertaken by sole individuals. The crucial prerequisite is the use of the open call format and the large network of potential laborers.*" A few months ago, he also published an article in Wired to describe "The Rise of Crowdsourcing" (Howe 2006b).
  Crowdsourcing is a process of receiving work or funding from a crowd of people, generally with the use of online tools. The term results from the combination of two words: *crowd* and *outsourcing*, i.e. the notion of this term is that a task or a work is undertaken, and it is outsourced to the crowd. The inception of crowdsourcing is that more hands working towards the same task are better than a couple of hands. In addition to this, by attracting the participation of a large crowd of people, not only the quantity but also the quality is expected to be much better. The necessary data/information quality control is important in some cases, depending the targeting of the project.
  The most famous example from universal crowdsourcing projects or initiatives is Wikipedia. Instead of creating a regular encyclopedia by hiring the writers and the editors, Wikipedia offers the crowd the opportunity to bring the content into existence, on their own resources. It is written collaboratively by anonymous persons who freely offer their services and contribute by providing content without any payment. The users of Wikipedia should be abreast of the fact that all articles have not the same quality, but they may contain disputable information or not according to the truth (Wikipedia 2018).

One of the most recent crowdsourcing initiatives in the cultural heritage domain is Rekrei, which means recreate in Esperanto, a language that was built for the purpose of international universality. It is a crowdsourced project to collect photographs and images of different cultural heritage objects such as monuments and artefacts devastated by natural disasters or human intervention. The project, which is a global initiative, aspires to use these data to create *3D* representations and thus, to contribute and help in preserving the global, shared, human cultural

Fig. 3.9 Tablet Nirgul: 3D reconstruction model (Mosul Museum). Created by Petr Vavrecka. (*Source* Rekrei)

heritage. By using crowdsourced photographs and images from any holder, tourist or expert, and by applying photogrammetric techniques, Rekrei generates *3D* representations, contributing to the preservation of memory of cultural heritage that has been lost. Initially, Rekrei was called Project Mosul and founded by Matthew Vincent and Chance Coughenour. They started this initiative after the destruction of cultural heritage in northern Iraq to create a network of volunteers across the globe to support in the digital restoration of the cultural heritage under risk or that has been lost (Vincent and Coughenour 2018). An indicative example from Rekrei gallery is given in Fig. 3.9 where Tablet Nirgul from the Mosul Museum was *3D* reconstructed.

Another real (photogrammetric) example from crowdsourcing campaigns in the cultural heritage domain is given in one of the following sections (Sect. 3.3.2). In this case, a designated monument, namely Plaka bridge in Greece, collapsed due to the extreme weather conditions in the area. A crowdsourcing campaign was initiated to collect any type of images that could be used for the *3D* virtual reconstruction of the bridge.

- **Libraries**

  Libraries, which could be either conventional or digital or both, are an incredible source of data and information on cultural heritage with an amazing toolbox containing maps, drawings, photographs, films, historic documents, etc. The reason is quite simple. The presence of libraries with their classical mode, full of books, documents and other paper-type stuff, for hundreds of years, provided the opportunity to gather enormous volumes of data and information. With the explosion of ICT and Internet technologies, nearly all libraries began the transition to the digital age, and their transformation to digital and on-line libraries. Apart from digital preservation, at the same time, the libraries preserving the archival library material since this is the real tangible and primary stuff. With digitization processes and internet technologies, they have created huge digital infrastructures to host this cultural heritage treasure and offer the on-line material of the library across the world.

The examples of (digital) libraries are many in every single part across the world, from east to west and from north to south. Let's take an example to look at the respective capabilities and sizes offered to the visitors and users; the example of the New York Public Library (NYPL). The NYPL has been the main supplier of free books, information, ideas, and education for all New Yorkers since 1895. It is the United States' (US) largest public library system and has as a prominent attribute the combination of the 88 neighborhood (library) branches and four scholarly research centers. The NYPL is unique and collects a remarkable wealth of resources and cultural heritage material of high importance, for architecture, history and preservation. The library presents excellent statistics, as illustrated in Fig. 3.10 (data from the financial year 2017), while its extremely important archival material in the area of topographic maps, photographs, old designs, etc., is a valuable assistance for researchers in the field of cultural heritage preservation. A topographic map of 1906 (Fig. 3.11) shows the street system and grades of that portion of the second ward (Town of Newtown), Borough of Queens, City of New York, as retrieved from the NYPL digital collections (NYPL 2018).

However, libraries are not only digital or physical infrastructures that host important and rare archival material. At the same time, they are also institutions with rich educational and social work activities and a wide grid of research initiatives. The NYPL's neighborhood libraries in the Bronx, Manhattan, and Staten Island are being remodeled into centers of educational innovation and service and pivotal community hubs that offer far more than just free librarian documents and other stuff. The NYPL's local libraries are engaged and actively contributing in closing the digital gap; one out of three dwellers in New York does not have Internet access at home. The students in the New York City public schools do have access in NYPL to support their homework. In addition to that, the immigrant communities in New York City rely on the classes organized from the NYPL for English language and literacy, while job seekers hinge on the resources provided by the NYPL (2018).

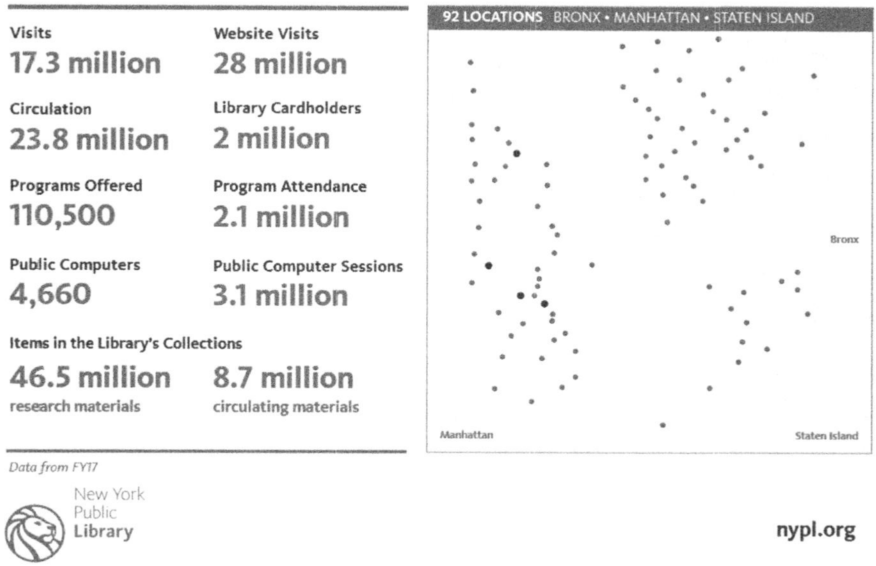

Fig. 3.10 The New York Public Library at a Glance (data from the financial year 2017). (*Source* NYPL)

- **Museums—Collections—Inventories**
  ICOM (2018a) is an international organization founded in 1946 by and for museum professionals. It is a unique network of more than 37,000 members and museum professionals representing the museums' community worldwide. According to ICOM, "*a museum is a non-profit, permanent institution in the service of society and its development, open to the public, which acquires, conserves, researches, communicates and exhibits the tangible and intangible heritage of humanity and its environment for the purposes of education, study and enjoyment.*"
  In 1986, ICOM adopted and introduced the Code of Professional Ethics, and in 2001, it was amended and renamed to ICOM Code of Ethics for Museums. The last revision took place in 2004, and in this document the minimum standards of professional experience and implementation for museums and their staff are set. It is considered as the cornerstone of ICOM, and none can join ICOM without accepting this Code of Ethics for Museums. This Code contains some principles for the museums (ICOM 2018b):

  – *Preserve, interpret and promote the natural and cultural inheritance of humanity.*
  – *Maintain collections; hold them in trust for the benefit of society and its development.*
  – *Hold primary evidence for establishing and furthering knowledge.*
  – *Provide opportunities for the appreciation, understanding and promotion of the natural and cultural heritage.*
  – *Hold resources that provide opportunities for other public services and benefits.*

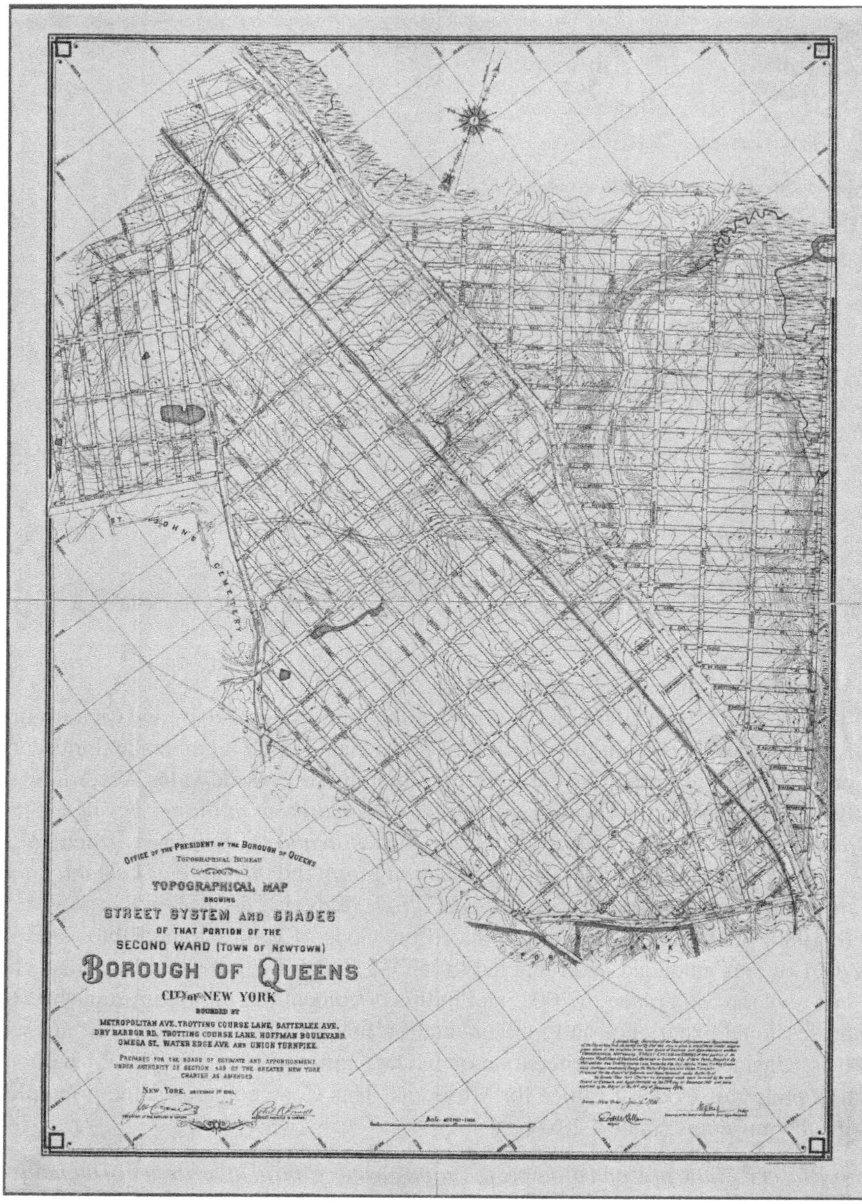

Fig. 3.11 Lionel Pincus and Princess Firyal Map Division, The New York Public Library. (1906). Queens Borough, Topographical Bureau. Topographic map showing street system and grades of that portion of the second ward (Town of Newtown), Borough of Queens, City of New York. Retrieved from http://digitalcollections.nypl.org/items/d7defd10-0dbb-0131-2856-58d385a7b928

- *Work in close collaboration with the communities from which their collections originate as well as those they serve.*

– *Operate in a legal manner.*
– *Operate in a professional manner.*

Museums are one of the most important pillars of the cultural heritage ecosystem. The museums are morally bound to act for their tangible and intangible character. Primarily, they are responsible for cultural heritage protection and promotion. As they have the legal obligation to collect, preserve and promote their collections, which are outstanding public legacy, their contribution is unique. They possess valuable information for protecting cultural heritage from harm or any loss, which should be exploited for the benefit of its safeguard. Consequently, museums have particular responsibilities to people and societies, for the provision of the necessary care, but also for the accessibility and interpretation of primary information brought together and kept in their collections.

The pivotal role of inventories and collections in cultural heritage documentation and management has long been accepted. They are absolutely necessary, for many purposes; historical, architectural, technical. For example, the Louvre Museum (Louvre 2018), the world's largest art museum and a historic monument in Paris, France, holds an incredible number of objects covering the period from prehistory to nowadays. The museum maintains various databases, collections and inventories, which are accessible through the web, providing imaging and descriptive information.

Inventories are also very important for cultural heritage. A well-known and open source inventory example is Arches. It is an open-source, geospatially-enabled software platform for cultural heritage inventory and management, elaborated jointly by the GCI and World Monuments Fund (WMF). The system is freely available for organizations worldwide to download, install, and configure in a manner conforming with their individual requirements and needs, and without restrictions (Arches 2019). According to Myers et al. (2012), Arches incorporates widely adopted standards (for heritage inventories, heritage data, and information technology) so the kernel offers a solid base that everyone interested in, be able to customize to specific needs. As an open source system, Arches is available for free, and allows adopters to share their resources to further improve it in mutually beneficial ways as well as maintain it.

- **Platforms**

  Platforms, which are essentially digital, help the creators to organize, store, maintain and disseminate data, almost anywhere on earth, even in the most remote parts of the world with an internet connection. One of the biggest projects worldwide is Europeana.

  Europeana is the digital platform of the EU for cultural heritage. It is a single digital access point to millions of books, paintings, films, museum objects and archival records that have been digitized across Europe. But what is the origin of this European initiative? GABRIEL, a European Commission (EC) funded project, was the first project that joined 43 national libraries. It was stopped as a separate webservice on June 2005 while its contents have been incorporated into the www.theeuropeanlibrary.org. Three more projects, TEL-ME-MOR, TELPlus and

FUMAGABA broadened "*The European Library*" with more national libraries and greater standardization and search capabilities. In 2005, six heads of European States, namely, French President Jacques Chirac, German Chancellor Gerhard Schröder, Italian Prime Minister Silvio Berlusconi, Spanish Prime Minister José Luis Rodriguez Zapatero, Polish President Aleksander Kwaśniewski, and Hungarian Prime Minister Ferenc Gyurcsány, signed a letter and asked EU officials to support the project. As a result, the national libraries of 19 European nations agreed to support the project. On 7 July 2005, José Manuel Barroso, President of the EC, replied to the letter and welcomed this initiative. By 30 September 2005, the EC adopted a strategy called "*i2010: Digital Libraries strategy*". This strategy described the digital libraries initiative vision and three key areas for action: digitization of analogue collections, online accessibility, and preservation and storage. Europeana was launched in 2008 (Europeana 2018).
As (Aparac-Jelušić 2017) argues, the foundation of Europeana as the unique access point to European cultural heritage is a relevant achievement. The provided digital content and services, and slowly but certainly raised public awareness about the power of digital media. Apart from that, the digitization of the European cultural heritage has significantly meliorated the accessibility of cultural heritage stuff for research, education, culture, and pleasure. It is evident that Europeana promotes the European culture, while at the same time, EU as founder and co-financing authority will continue to improve Europeana services and performances towards a sustainable and user-oriented future.

3.2.1.2 What Type of Data?

In addition to the (new) data and information that are collected during the survey and recording phase, e.g. for the documentation of a historic building, inevitably and it is fortunate that the documentation process unites and incorporates a set of other important, historic and archival, features, data and information. In this case, the question raised is: what type of data are of interest to the recording and documentation projects of cultural heritage?

So, what type of data? What are the ranging of forms, and categories for the requested information and data, in cultural heritage recording and documentation? The ensuing catalogue of potential data types is providing, at the same time, the eventual boundaries where a documentation team should search. For each one of the reported data types a short description is specified to give an exposure of each. The typical examples provided are serving as an indication to underline the state of their great significance.

- **Descriptions on historic texts**
  The ancient and historical texts are unique sources in cultural heritage documentation. Among these documents one can find outstanding value of the oldest manuscripts on papyrus, ancient hand-written texts, or even more recent technical or historiographical texts of the last centuries or decades. Unquestionably, this is a salient source of information, which may contain descriptive, historic or even

artistic and technical details, which are extremely useful in cultural heritage documentation research. The detailed examination and interpretation of these sources may reveal an astounding wealth of information, but it should be made in a scientific and systematic manner and always in combination with other sources, in order to advance meaningful and solid conclusions.
The interpretation of ancient and historic texts to preserving cultural heritage, the creation of digitized inventories, libraries and databases of cultural heritage records, imaging and information technology have played an essential role in not only preserving the historic character of these documents, but also in making these documents accessible, animated and alive.

- **Drawings and sketches**
Historic and old drawings and sketches are supplementary, recognized and salient sources of information in the steps taken in order to achieve recording and documentation of cultural heritage. Especially, the (metric) designs which, under normal conditions, have the requisite precision, indicated by the numeric or graphic scale, may prove to worth a great focus as tools during the documentation of a historic building. The usage of such sources of information may be evidence-based and efficient for many reasons, three of which are mentioned in the following:

  - This source may come up with unique information that at the present time is not able to be obtained, and this is of great importance to be in the know about the evolution of the presented object.
  - It can be used as the foundation to compare the condition of the presented object, between different or more epochs, in a change detection approach.
  - This archival and tangible stuff, to some extent, will be part of the documentation archive, available for any future use at all times.

One such case of using historic drawings in a *3D* survey is the work of Canciani and Saccone (2011) in the church of St. Thomas of Villanova in Castel Gandolfo, in Italy. The church was designed by Gian Lorenzo Bernini and built between 1658 and 1662, while it was commissioned by Pope Alexander VII Chigi. The study concerned an integrated survey eventuated in the church and a *3D* modelling and reconstruction of all the complex elements and the hard to measure geometries. The aim of this study was to compare the *3D* survey model of the church with the work of Bernini to come into a more solid perspective for the design and the planning methods used by Bernini. A representative illustration from this work is depicted in Fig. 3.12, where the outcomes of the *3D* survey are superimposed in the historic drawings of 1658–1660.

Another example is illustrated in Fig. 3.13, which is a drawing designed by Vincenzo de' Rossi (Italian, 1525–1587) and acquired by The Metropolitan Museum of Art in 2013. The sheet, which depicts a comprehensive design for an altar and is inscribed and signed by the artist at bottom right "Vincentio Rossi", can be considered the first genuine architectural drawing. It is nearly certain that it is connected to an early and prestigious commission for the altar of the Confraternita dei Virtuosi in the Pantheon, Rome, commissioned from the artist in 1546.

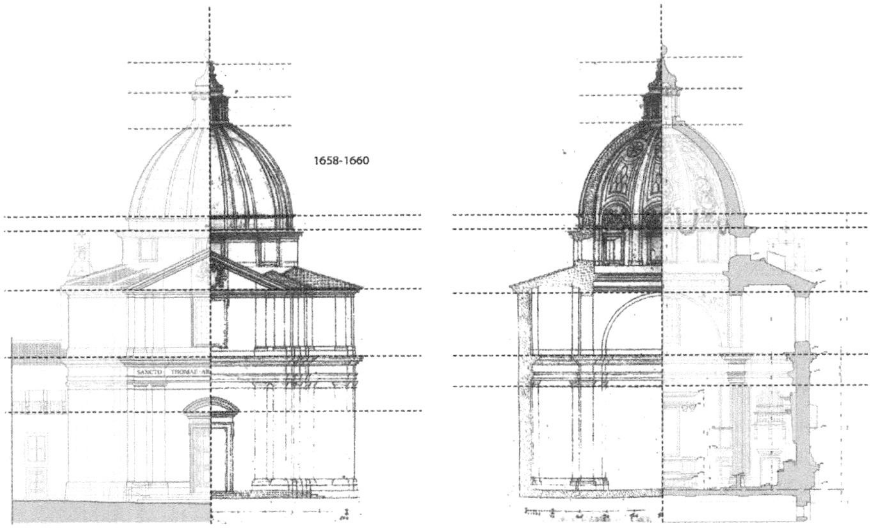

Fig. 3.12 Comparison of survey drawings (2010) and Bernini's design drawings (1658–1660) in the Church of Saint Thomas of Villanova in Castel Gandolfo in Italy (Canciani and Saccone 2011)

The drawing constitutes a nearly ideal case of an orthographic representation, combining a frontal elevation, floor plan, side view, and section of an altar (MET 2019).

- **Maps**

  According to the International Cartographic Association (ICA), "*a map is a symbolised representation of geographical reality, representing selected features or characteristics, resulting from the creative effort of its author's execution of choices, and is designed for use when spatial relationships are of primary relevance*" (ICA 2018). However, maps are not important only as tools for representing our world nowadays. Maps, and especially the historic maps, illustrate the imprint of the world in the past decades or hundreds of years, and they retain vital information to help us understand the connection between the past, the present, and the future.

  The tremendous increasing of data availability is visible everywhere, in mapping, geospatial infrastructures and social media data, almost in everything that happens around the world. There is a need to visualize this data, and altogether comprise a massive pool of knowledge, which is requesting analysis and interpretation. This new data ecosystem raises awareness and needs for new methods, tools, and human resources to process, analyze and interpret this data into practical information which is able to provide new knowledge for the world. This is why big data is almost in every subject and is becoming ubiquitous. However, by nature, mapping was always a big data scientific activity, targeting different disciplines, from collection and processing, to analysis and representation.

Fig. 3.13 Design for an altar surmounted by a crucifix in four different views by Vincenzo de' Rossi (Italian, 1525–1587). (*Source* MET/CC0 1.0)

How about the past? How about what happened or changed decades or centuries ago? The imprint of the past can be done through historic maps. The features of the maps, their typology and symbology, but even more, the map content itself, is the actual characteristic of this historic and valuable visualized source of information. Every map reflects, to the representation of concrete spatial and content background, knowledge that is visible or kept out of sight, within the lines, points, symbols and colors, and inherited to the next generations.

Undoubtedly, historic maps contain cognitive, graphical and descriptive features that can be further exploited with the use of GIS and ICT-based tools. The powerful fusion of heterogeneous cartographic and other data (e.g. from art and history), can lead to the generation of new knowledge. It can also drive the development of new tools that are capable of supplying research with new multiplicative and complex representations by merging visual and non-visual information. In addition to that, pairing and comparing data from separate sources, may lead to geometric and descriptive information that is usable for the documentation and management of cultural heritage objects such as historical buildings.

The historic map of Athens, illustrated in Fig. 3.14, is a typical example on how historic maps can contribute to the research and also to the cultural heritage documentation and management of monuments and historic buildings. It is very well known that Athens is one of the most important cultural heritage places in the world, not only because of the Acropolis and Parthenon, but also because it symbolizes the cradle of Western civilization and the birthplace of democracy. In this map, historic buildings and monuments, buildings and building blocks, landscape information, contour lines, road network, names and toponyms, and many more are rendered in the scale of 1:12.500. This map is originated back to the 1870s, and it was created by J. A. Kaupert.

- **Photographs**

  The capacity of the old photographs in the documentation of monuments and sites is a great and irreplaceable source of information in many cases. All these decades, since the photo was discovered, old photographs are playing a unique role in the documentation process, as it is the only true and priceless visual representation of the past. This is something that cannot be offered by other means, such as drawings or sketches.

  Historic photographs have high level density of information. They constitute a tangible source of information for architectural analysis, building modelling, condition analysis in monuments, sites and many more cultural heritage objects. In that respect, it is so important that there is possible to *2D/3D* reconstruct geometric information of the objects illustrated in the old photographs. It is also possible to use photogrammetric methods and calculate the position of the camera at the time of the recording, or even the unknown camera geometry. Nowadays, from the photogrammetric point of view, historic photographs are mainly digitized photographs, i.e. digital images used in photogrammetric workflows. Sometimes there is missing or minimum amount of information about the object of interest. Low radiometric and geometric resolution may also occur. But, in any case, historic photographs remain an invaluable treasure of information that the researchers must make the most of it.

The examples are hundreds or even thousands in different parts of the world, and concern different objects and types of cultural heritage. It is not workable to give many examples, but only just a few to demonstrate the importance of old and archival photographs in different projects and applications.

In Technical University of Berlin, Germany, where they have the privilege to have access in the Meydenbauer (Sect. 6.2) archive, (Wiedemann et al. 2000) discussed how photogrammetry can contribute to determine sufficient information by using historic images. For this reason, they used archival photographs from the Meydenbauer (Fig. 6.1) archive. They presented a summary of the data sources used in destroyed buildings in the city center of Berlin, the image orientation problems faced during the processing stages, and some restitution techniques to produce photogrammetric outcomes.

In 2003, (Grussenmeyer and Jasmine 2003) used several archive photographs of the Beaufort castle (South Lebanon), also called Qalaat el-Chaqif (12–17th century). The photographs, vertical and oblique overall views on glass plates, were acquired by the French army between 1935 and 1937. In this study, they also used the castle survey of 2002 and additional aerial oblique images taken from a helicopter. Terrestrial images were captured during the fieldwork and used as well. They used this archive documentation to perform a 3D restitution of the ruined and buried historic structures of the castle. This recording and documentation were necessary to support the consultants to establish an excavation framework and restoration tasks in the area.

Other studies experimented with the fusion of historic photographs with terrestrial laser scanning (TLS) data to perform *3D* reconstruction. For example, (Hanke et al. 2015) used two historic photographs (Fig. 3.15) from around the 1920's and *3D* documentation of the church from TLS to implement a photogrammetric reconstruction of the altar at the Capuchin monastery of Kitzbühel in Austria. In a similar project, (Bitelli et al. 2017) used historic photographs and TLS for the 3D virtual reconstruction of destroyed structures. More specifically, the performed the 3D reconstruction of the civic tower of the small town of Sant'Alberto, near the city of Ravenna in Italy. This tower was destroyed in December of 1944 by the German troops in retaliation when they were forced to leave the area.

3.2.2 Cultural Heritage and Open Data

The ideal circumstances for those working in the cultural heritage domain is to have on their disposal open data. Usually, not all data is freely available and open to any person interested in further usage and sharing. But what is open data? It is the data that can be used for free and distributed by anybody, at most, merely subject to the

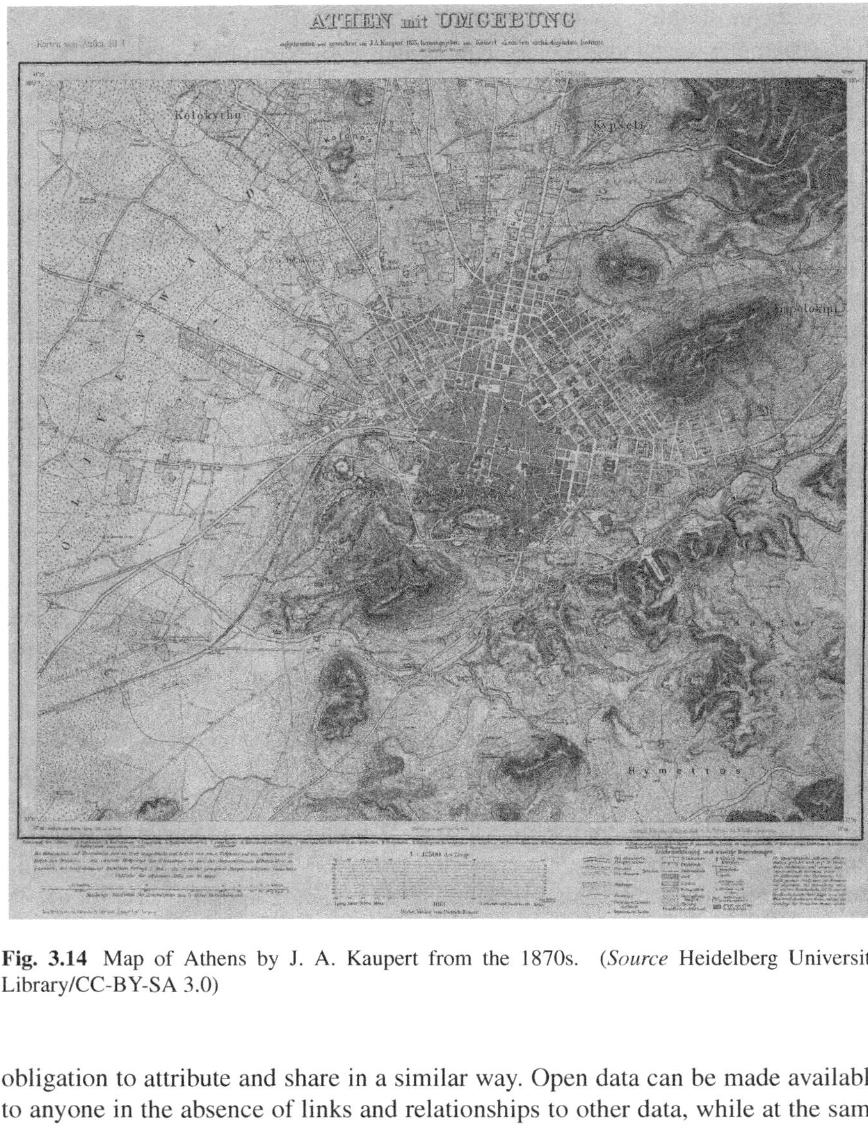

Fig. 3.14 Map of Athens by J. A. Kaupert from the 1870s. (*Source* Heidelberg University Library/CC-BY-SA 3.0)

obligation to attribute and share in a similar way. Open data can be made available to anyone in the absence of links and relationships to other data, while at the same time, data can be linked without being freely available for reuse and sharing among a number of recipients.

On the other hand, Linked Data is one of the core concepts and building blocks of the Semantic Web, i.e. an extension of the World Wide Web (WWW), that was invented by Sir Tim Berners-Lee in 1989 (Webfoundation 2018), which through standards by the World Wide Web Consortium (W3C) is making links between datasets comprehensible both to humans and to machines. Semantic Web is also known as the Web of Data or Data Web. In fact, for making those links, Linked Data

Fig. 3.15 Historical photograph showing the desired altar ensemble as it was before WWII in the Capuchin monastery of Kitzbühel in Austria (Hanke et al. 2015)

makes available for use the best practices, as a set of design principles for sharing machine-readable interlinked data on the Web (LinkedData 2018).

Open Data is not the same in value as Linked Data. What is Linked Open Data (LOD)? It is a powerful combination of Linked Data and Open Data, as it is linked and uses open data sources. The more entities such as objects, persons, etc. are connected together, the better for the Web of Data power. In order to attain this link and combine to form a single entity the enormous datasets from dissimilar data sources, a set of principles should be adopted.

In 2006, the person who invented the WWW, created and publicly supports the Semantic Web and Linked Data, Sir Tim Berners-Lee, introduced the four principles for LOD (Berners-Lee 2006).

1. Use Uniform Resource Identifier (URI) as names for things.
2. Use HyperText Transfer Protocol (HTTP) URIs so that people can look up those names.
3. When someone looks up a URI, provide useful information, using the standards (RDF, SPARQL).
4. Include links to other URIs so that they can discover more things.

A URI is a single global identity, acting like a unique ID, for all things linked. This singularity enables the users to recognize those things and combine them with others without uncertainty. In addition to that, with URI one thing from one dataset

Fig. 3.16 The 5-star rating system for LOD. (*Source* (Berners-Lee 2006))

is the identical to another in a separate dataset for the reason that they have one and the same URI.

The Resource Description Framework (RDF) is a standard model, developed by the W3C, publishing and interchanging data on the Web. It is the standard used in a semantic graph database, i.e. the technology developed for interlinked data storage, and it is also referred to as an RDF triplestore. The triplestore maps the miscellaneous relationships amid entities in graph databases.

SPARQL is an RDF query language and protocol developed by W3C. SPARQL can be used to define queries across different data sources, while the outcomes of SPARQL queries may be RDF graphs or results sets.

Sir Tim Berners-Lee proposed also a 5-star deployment scheme for LOD. The 5-star rating system begins with one star and data gets more stars when proprietary formats are disconnected from and links are added (Berners-Lee 2006). In fact, this idea put forward by Berners-Lee was also accompanied by a figure printed on a cup of coffee (Fig. 3.16). According to this approach, what does it take to be 'awarded' each of the five stars, and what are the benefits the users of those datasets draw from going up the stars count?

- **1-star open data**
  The 1-star open data is described exactly as data available on the web, regardless their format, but surely with an open license, so as to exist as open data. Everyone can view, search, store, modify, and share the data with anyone.
- **2-star open data**
  In order to shift from one to two stars, the open data should be available as machine-readable structured data. A very typical example is an Excel spreadsheet rather than a table in a raster format (image). Everyone using 2-star open data is free to do anything that does with a 1-star data, with the addition of directly processing

it with proprietary software and exporting it to any other structured format. The existence and dependency on the proprietary software make the data still locked.

- **3-star open data**
  If the proprietary software which is necessary for data analysis is removed and thereby every user does not require proprietary software, then the data are open with 3 stars. Following the previous example, imagine now, instead of providing an Excel spreadsheet, make available for use a comma-separated values (CSV) file that does store the tabular data in simple plain text.
- **4-star open data**
  The 4th star is assigned to data that uses open standards from W3C in order to identify things. As already mentioned, such standards are RDF and SPARQL. RDF is the standard used in a semantic graph database (RDF triplestore), while SPARQL is the W3C-standardized query language. For that reason, the 4th star is 'awarded' to data, because by representing data in a graph database, the user may link to it from any other site or reuse parts of the data.
- **5-star LOD**
  The 5-star data is not only open, but it is LOD. The 5-star LOD is open data available on the web linked to other data. With the support of the W3C standards and Linked Data principles, as already discussed, this linkage is feasible. In this case, data providers link the data to the other data in order to provide context, and the users of 5-star data can find more and more interlinked information during the time of using the data.

Linked Data removes the barriers between various sources and makes data integration and browsing easier, especially when dealing with complex data. The standards used are contributing amazingly while the guidelines allow for data models' updates and extensions. By data presentation in a linked mode and denoting a set of global principles, this surprisingly contributes in increasing data quality. Linked Data representation through a semantic graph database generates semantic links between various heterogenous data from different sources and deduces new knowledge.

Besides, there are further qualitative benefits from the usage of LOD. By linking open datasets from different sources, this leads to reinforcing creativity and cultivation of innovation to citizens, researchers, in academia and enterprises. Everyone can use all those datasets and create new applications and generate new knowledge, which again is for the benefit of all. LOD encourage the developers to develop new algorithms to come across new insights that may never have been considered. Undoubtedly, the common standards and principles and the open data policy for transparency make LOD very beneficial to organizations and society.

Especially for cultural heritage, as Marden et al. (2013) are underpinning, LOD offers a new course for cultural heritage institutions to share their valuable property with the society and the wider audience. This will be a game-changer in the traditional relationship between those who are holding the knowledge, knowledge interpreters, and the consumers of knowledge. With a powerful ecosystem based on multiple and heterogeneous datasets, the users with all levels of knowledge and skills are able to retrieve, access and analyze information. They believe that this new way to obtain, examine and interpret cultural heritage information might update the way cultural heritage is defined.

There are many different types of open data and many more ways to exploit open data in cultural heritage. One of these ways is the use of open imaging data to perform *3D* reconstruction and visualization. In his work, (Themistocleous 2017) used open data from social media videos to perform *3D* model reconstruction and visualization of a cultural heritage site. Structure from motion (Sfm) techniques were used for extracting and building the *3D* model of a cultural heritage monument in Cyprus. In practice, the author used a video from YouTube to create a quality geo-referenced DSM, by using images extracted from a low-resolution video for visualization purposes.

3.2.3 Intellectual Property on Cultural Heritage

According to the World Intellectual Property Organization (WIPO), "*Intellectual property refers to creations of the mind: inventions; literary and artistic works; and symbols, names and images used in commerce. Intellectual property is divided into two categories: (a) Industry Property includes patents for inventions, trademarks, industrial designs and geographical indications; (b) Copyright covers literary works (such as novels, poems and plays), films, music, artistic works (e.g., drawings, paintings, photographs and sculptures) and architectural design. Rights related to copyright include those of performing artists in their performances, producers of phonograms in their recordings, and broadcasters in their radio and television programs*" (WIPO 2018d).

At the same time, as the greater number of cultural heritage objects and collections hosted in the cultural institutions are in the public domain, there is also a big number of objects and collections with Intellectual Property Rights (IPRs) attached to them. In most countries, after a period of 70 years, starting from the death of the author/creator, the object/collection becomes part of the public domain. This means that from that moment on, any person is free to use the work/object/collection. One more variable that affects the IPRs of an object/collection is the fact that third-party rights may attached to this. These two conditions are induced where an object/collection will be considered as part of the public domain.

The activities of cultural heritage institutions are targeting the management and preservation of cultural heritage property. The greater number of such materials seems to be under the copyright domain. Different types of materials such as photographs, images, drawings, and many more are fall into the WIPO definition, as provided above, and consequently, such materials are under the copyright protection. The IPRs should be always considered, whether they exist or not. Just to give an example, imagine a collection that is digitized to be made accessible over the Web to millions of potential users.

The cultural institutions interested in the digitization or licensing for reuse cultural heritage information need to consider where a specific cultural heritage object is protected by copyright or by any other related right, or it is under the public domain. During the digitization process, new rights may emerge, for example, the IPRs may arise from third parties in the scanning process. And it is not only the process of scanning but also other neighboring and relevant processes such as filming or photographing, indexing of digital cultural heritage objects, annotating, semantic tagging of the metadata, even database rights, or any other rights that may applied case by case. Therefore, the cultural institutions that may wish to release any type of cultural heritage information for reuse, at the same time, they should define who is the owner of any such rights that potentially will be created. In addition to that, they should define how third parties that may be involved in the process are able or not and how to further exploit them.

3.2.3.1 Regulatory Framework and International Treaties

The rights of the authors and creators are mentioned in the international regulation framework but also in the national legislations. Globalization, the boom of the ICT sector, including the Internet and social media and networks, has an impact and inevitably affected the developments on the cultural heritage property management sector. Even though copyright is a national matter, there are also cross-border issues and thus the authors' and creators' rights cannot be handled within the national borders.

Various international organizations are acting in the IPR's domain. UNESCO (UNESCO 2005) through the "Convention on the Protection and Promotion of the Diversity of Cultural Expressions" (2005) is one example. WIPO (WIPO 2018d) is another example of an international organization that promotes the IPRs protection by initiating the cooperation between countries.

In 1883 was the first global agreement on the regulation of intellectual property, known as the Paris Convention for the Protection of Industrial Property (WIPO 1883a). This convention, which was amended in 1979, aimed at the regulation of industrial property in a wider sense, extending from patents, industrial designs, etc. Three years later, in 1886, the Berne Convention for the Protection of Literary and Artistic Works (WIPO 1886b), which also amended in 1979, is the Convention, which dealt with the protection of moral rights. After many years, in 1961 a new convention came to the fore, the Rome Convention for the Protection of Performers, Producers of Phonograms and Broadcasting Organizations (WIPO 1961c), which widened, for the first time, the scope of protection to neighboring rights.

European Union supports that an organized and efficient intellectual property infrastructure is required to make certain that the stimulation of investment in innovation and to keep away IPRs violations which reflect economic damages. For this reason, EU is targeting to secure that this infrastructure permits the creators and inventors in its borders, to gather the proper returns. Legal instruments, such as the Directive on the enforcement of IPRs, known as IPRED, which was adopted in 2004, obviate the IPRs infraction. The Directive demands from all the EU countries to introduce specific measures against those having an illegal engagement with IPRs (EC 2018).

3.2.3.2 What are the Types of Rights?

The copyright regulation is part of national regulatory framework. However, even though there are different perspectives amid the nations worldwide, the national regulations are becoming more consistent, extending the power of international treaties, conventions or even European directives, in the case of European countries. In this context, the copyright owner is the person who created the work. The copyright holder is enabled with full control over the use of this specific work and therefore financial benefits may be assumed in the case of further usage of this work. The exploitation of the copyright holder's work directly imposes a financial compensation to the copyright owner. Besides, the owner has the privilege to consent for the use of the work under specific agreement, without having a compensation payment.

There are different types of rights for the copyright holders, in return to the consideration of their work. In addition to that, the owners have the right to offer their work free or under an internationally accepted license scheme such as Creative Commons (CreativeCommons 2018).

- *Economic rights*: the owners have a financial compensation for their investment in creating a specific work. The economic rights are transferable. The economic rights allocated into reproduction rights (Art. 2 of Directive 2001/29/EC) and the right of communication to the public (Art. 3 Directive 2001/29/EC).
- *Moral rights*: these are the rights that safeguard the relationship amid the owner and the work that created. The moral rights are not transferable. As they have a personal dimension, the owner/creator has a right of paternity/maternity, the right to integrity, and also the right to bring or not the work to the market or make it publicly available.

The IPR issues are complicated issues, and legal advice is always necessary for safeguarding and resolving the position of every person or institution involved.

3.2.4 Sensors and Systems

Both sensors and systems reflect to a huge number of potential applications and choices in culture heritage documentation domain. Sensors and systems are those tools that contribution to data acquisition so as to proceed with data fusion, i.e. merging data and knowledge originating from different sources.

The constant development of new sensors and systems for data acquisition and for monitoring purposes in cultural heritage is so obvious nowadays. Many and different solutions in terms of applicability, size and cost are arriving in the market continuously. The importance of cultural heritage recording and documentation in mapping and preservation, and the tremendous developments in the ICT and sensor sector, highly contribute in having this growth day by day in international levels.

The generation of 2D and 3D outcomes in cultural heritage domain presupposes the collection of data by using the appropriate type of equipment. Nowadays, a large number of sensors and systems are available for use in digital recording and documentation of cultural heritage. It is not possible to analyze every single solution, but this section will illuminate the available options. The following 5 categories of sensors and systems could be assumed in this clustering:

1. Image Assisted Total Stations
2. Passive sensors
3. Active sensors
4. Mobile mapping systems
5. UAVs

3.2.4.1 Image Assisted Total Stations

Image Assisted Total Stations (IATS) are geodetic systems that consolidate the precision of total (geodetic) stations, in addition to the image acquisition mode, in one single integrated system. One or more cameras are integrated as part of IATS and enable the system to deliver georeferenced and oriented images in an accurate and efficient way. In addition to that, by applying image processing and image recognition techniques, in combination with other surveying techniques, new measurement approaches can be achieved to provide supplementary outcomes.

In the recent years, almost all major total stations' commercial manufacturers had an interest to develop and merchandize image assisted systems. For this reason, they released total stations that embody at least one camera, commercially known as Image Assisted Total Stations or simply IATS. Such a system is able to acquire images and live video stream, which is absolutely oriented in the selected coordinate system. The exterior orientation of the images is directly known, due to the IATS known station, orientation and mounting offsets. The images and the video can be used in many ways, depending the user's needs and the system's functionalities. For example, some manufacturers provide supplementary software that allows the user to perform photogrammetric post-processing. In the case the system has a scanning functionality available, a real scanning of selected areas is feasible through the IATS.

In Fig. 3.17, 4 different total stations with imaging and scanning capabilities are presented. First is the Trimble SX10 Scanning Total Station (Fig. 3.17a, which is a full high-accuracy total station enabled with a high-precision 3D scanner in a single system (Trimble 2018). Second is the Leica system, called Leica Nova MS60 (Fig. 3.17b), which is a MultiStation combining all available measurement technologies in one instrument (Leica 2018). The Topcon system of IS-3 series, which is illustrated in Fig. 3.17c is an advanced imaging and long range scanning robotic total station (Topcon 2018), while the last IATS system is the Pentax one from Series R-

(a) (b) (c) (d)

Fig. 3.17 Imaging and scanning total stations by Trimble (**a**), Leica (**b**), Topcon (**c**) and Pentax (**d**). (*Source* the respective companies)

Table 3.2 Basic technical specifications of the four imaging and scanning systems presented in Fig. 3.17. (*Source* the respective companies Trimble, Leica, Topcon, Pentax)

–	Trimble	Leica	Topcon	Pentax
Angle accuracy	1"	1"	1"	3"
Distance accuracy (1p)	1 mm + 1.5 ppm	1 mm + 1.5 ppm	±(2 mm + 2 ppm x D)	±(3 mm + 2 ppm x D)
Range (1p)	1–5.500 m	1.5 m to >10.000 m	3.000 m	1.5–5.500 m
Vision	Multi-camera	Single camera	Twin digital camera	Single camera
Scanning specs	6 min, Standard scan; 90° x 45°(H x V), Density: 0.5 mrad, 25 mm spacing @ 50 m	1 Hz mode, range noise 0.6 mm at 50 m	20 points per second	–
Scanning range	600 m	1000 m	2000 m	–

400VDN (Fig. 3.17d). The Pentax system is a reflectorless total station with image capturing capabilities (Pentax 2018) (Table 3.2).

3.2.4.2 Passive Sensors

Passive sensors are the one version of the non-invasive optical recording sensors. The other is the active sensors, which is discussed in the next section. Terrestrial active and passive sensors are not used only in cultural heritage but also in many other application areas in order to derive *3D* shapes. Many times, they are called as 3D imaging techniques and can be applied in architectural, industrial, medical and many other fields (Sansoni et al. 2009).

(a) (b) (c) (d)

Fig. 3.18 Conventional cameras by Canon (**a**) and Nikon (**b**), and thermal cameras by Flir (**c**) and (**d**). (*Source* the respective companies)

In general, passive sensor technologies capture target data through the detection of heat, light, radiation, vibrations or other phenomena occurring in the built and natural environment.

Examples of passive sensor-based technologies include: chemical, infrared, photographic and seismic. In any case, photographic is considered as the most known and useable technology. However, there are cases with some sensors in dual role; infrared and seismic sensors incur in both active and passive modalities.

The sensors can be mounted in different platforms determined by the application itself. Depending to what is being sensed by these devices, these various sensors might be mounted to airplanes, balloons, helicopters, kites, UAVs, vehicles or even in the space in satellites, or in the sea in boats even in submarines. In practice, it might be mounted to anything that provides a convenient point of observation. The data captured by remote sensing technologies is used for many and different purposes, from mapping to chemical measurements.

There are many examples from passive sensors used in the recording and documentation of cultural heritage. Indicative examples from the photographic and thermal type are illustrated in Fig. 3.18.

3.2.4.3 Active Sensors

An active sensor is a sensing device that needs for the particular purpose of functioning, an external source of power. In fact, it is the "opposite" of the passive sensors which in a direct way detect and make a response to some type of feed in from the physical environment. In this context, an active sensor is a device with a transmitter. The device transmits a signal, light wavelength or electrons to be bounced off an object, and the data are captured by the sensor upon their reflection to the surface of the object.

It is obvious that active and passive sensors have benefits and drawbacks. The passive sensor technologies cannot be detected by observed parties as they simply sense what is in the environment rather than depending on a transmitter whose operation might be identified with equipment. However, active sensors can be used when passive sources of observations by sensor are seemingly impossible task. For example, this could occur when the observed objects are not disposable to a camera during the absence of light conditions. Active sensor technologies like Light Detection and Ranging (LiDAR) can be used independent of daylight as they possess their own radiations on which to ground on their observations.

Examples of active sensor-based technologies include: GNSS, infrared, LiDAR, radar, seismic, sonar and x-ray. The active sensors can be expressed also as optical range or range-based sensors. In the context of this book, the active sensors can be distinguished into 3 main and widely used categories in the terrestrial domain:

1. Triangulation-based sensors
2. Time-of-Flight (ToF), pulse sensors
3. Phase-based sensors

The range-based sensors are extensively used for *3D* surveying and modelling applications, because they are capable of acquiring directly the *3D* geometry of objects and produce *3D* digital presentations through the point clouds and the range maps. This type of instrument captivated many users and professionals in the cultural heritage domain, working in the field of recording and documentation. Despite their high cost of purchasing such a system, terrestrial optical range sensors can effectively work from a few centimeters to a few kilometers, depending the technical specifications of the equipment itself, the object under study and the overall mapping conditions. The delivered *3D* data holds an accuracy, varies from few microns to few millimeters.

There is a wide range of *3D* measurement techniques which are described by the scale at which they could be used and by the number of measurements they could capture. Laser scanning can be used in the case the question is to provide a huge number of measurements for such like object sizes, while it is appropriate particularly for more complex objects. It enables the collection of an extremely large *3D* datasets in a short period of time. Apart from terrestrial use, laser scanning may also be deployed from the air, for instance by using a UAV. Laser scanning or *3D* scanning, as a surface-based *3D* measurement technique can result huge amount of points recorded in a systematic way, called point cloud, which after processing can lead to *2D* CAD drawings and *3D* CAD/BIM models or even to *3D* surface models, photorealistic textures and video animations (Boehler and Marbs 2004). According to Grussenmeyer et al. (2016), laser scanning is "*an active, fast and automatic acquisition technique using laser light for measuring, without any contact and in a dense regular pattern, 3D coordinates of points on surfaces*".

Regardless of the system used, laser scanning generates a point cloud, i.e. a point dataset of *XYZ* coordinates in a coordinate system, that depicts a spatial conception of the object under study. Additional information such as the Red-Green-Blue (RGB) values of the points may be also included in the *3D* dataset. In most cases, a point cloud is chosen to contain an extremely large number of *3D* points, i.e. a dense spatial object representation rather than a sparse point cloud.

Active range sensors based on triangulation

The laser scanners which are based on the triangulation principle of measurement are disposable in several forms, configurations and weights, like the following:

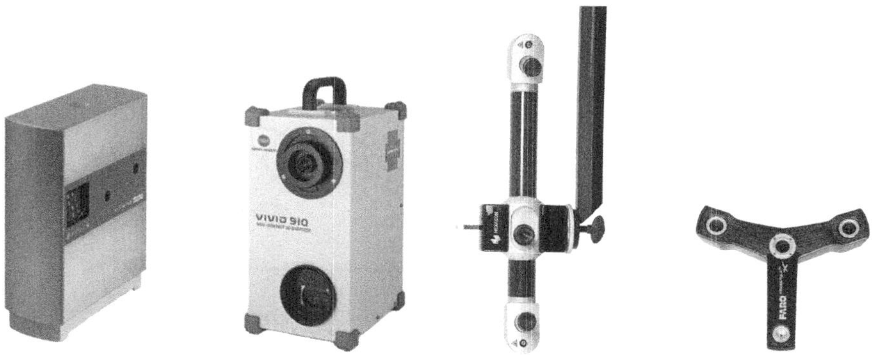

(a) (b) (c) (d)

Fig. 3.19 Triangulation and structure lights scanners from NextEngine (**a**), Konica Minolta (**b**), Hexagon (**c**) and Faro (**d**). (*Source* the respective companies)

- Handheld scanners, which are operating in close range tasks, or others are used as mobile devices.
- Scanners that are used over a tripod.
- Hooked-up scanners in mechanical arms.
- Static scanners which are used in combination with a turnable disk to carry the (small) object under scan.

To some extent, in these scanners they do not use laser as the main source, but as an alternative, they used white light, and they are called structured-light scanners (SLS). In this case, the light is projected in a structured pattern of either stripes or grids, however, they are operating based on triangulation. SLS are workable in a light-controlled environment for the reason that ambient light affects the measurements.

It is not possible to present all the available commercial solutions in the market from this category. In addition, the solutions are changing, and new products are constantly coming to the market. Nevertheless, four indicative examples are presented, two from triangulation scanners (Fig. 3.19a, b) and two from SLS (Fig. 3.19c, d). NextEngine is a triangulation scanner, developed by NextEngine Inc., a company based in Santa Monica, California, US. The cost is relatively regular to low, while its resolution, according to the developers, is 0.1 mm, and alike the error (NextEngine 2018). The Vivid 910 *3D* Scanner from Konica Minolta is also a triangulation scanner, and it has three types of lenses: a wide-angle, a middle-angle and a telescopic. The manufacturer provides specific instruction on the suitable lens to be used each time, based on the size of the object and the distance (scanner-object). The accuracy varies from 0.16–1.04 mm (*XYZ*), while the precision from 8–32 μm, both depending the lens used (Konica-Minolta 2018). The AICON SmartScan by Hexagon is an optical scanner based on structured-light technology. This 4 kg scanner is equipped with 5 or 8 megapixel digital camera, while the measurement fields vary from 60 mm up to 1.550 mm by changing the camera lenses and the base length. The working

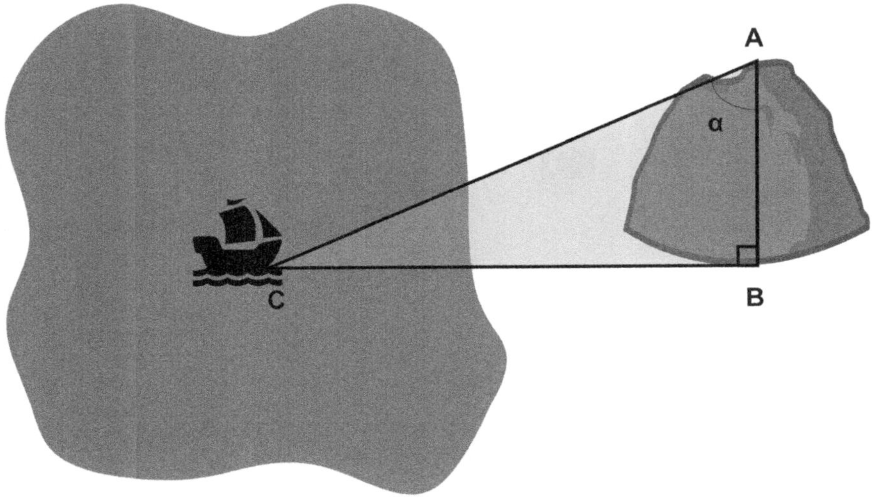

Fig. 3.20 Representation of triangulation as applied by Thales in 600 BC

distance extend between the limits 340–1.500 mm according to the configuration. The measurement volume ranges from a few millimeters to about one meter, and objects are digitized within seconds (Hexagon 2018). The last model is the FARO Scanner Freestyle3D. It is a handheld *3D* scanner and the manufacturer argues that this scanner is offering the largest scan volume in the market. According to the technical specifications of this scanner, for measurements within 1m distance, the user can succeed a *3D* point accuracy of even less than 1 mm, while the indoor scanning volume can reach up to 8 m^3. The scanner is operable in the temperature range of 0–40°C (Faro 2018).

In triangulation, the use of a triangle to calculate a distance is marked back to antiquity, and this formation is dated back to 600BC when Thales, a pre-Socratic Greek philosopher, mathematician, and astronomer, who is one of the seven sages of Greece, measured the distance of a ship, from one side and two angles of a right-angle triangle, as illustrated in Eq. 3.1 and also in Fig. 3.20.

$$BC = AB \cdot \tan \alpha \tag{3.1}$$

One of the most common principles of *3D* scanning is laser triangulation. It is simple and robust. In laser triangulation, laser technology can be used, and it is projected over an object. The image of the backscattered beam is captured by a camera and the line joining the laser, the object and the camera forms a triangle ($\triangle ACD$), as shown in Fig. 3.21, hence the term triangulation for the overall methodology. In other words, a triangulation scanner calculates the *3D* coordinates of an object by triangulating the position of a spot or stripe of laser light.

In Fig. 3.21, the following quantities should be considered:

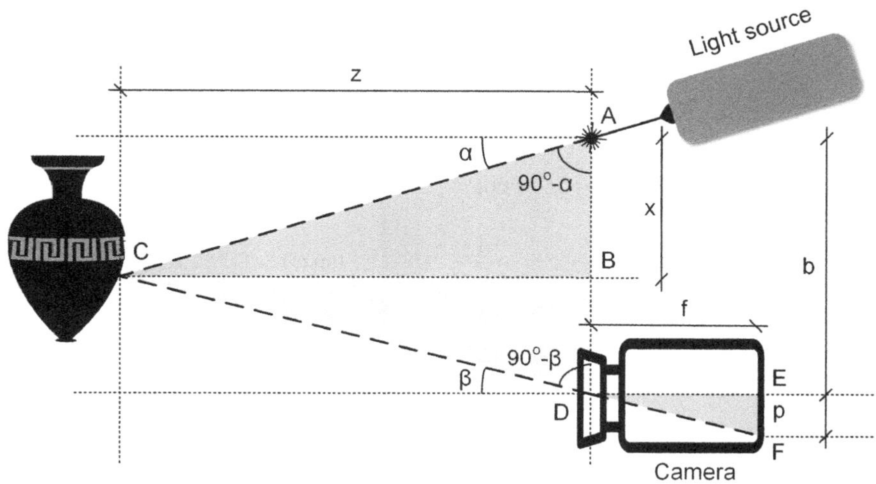

Fig. 3.21 Triangulation principle in laser scanning

f	Camera focal length.
b	Baseline (light source—camera).
α, β	The angles from the incident and backscattered beam respectively.
p	The position of the backscattered beam in the camera sensor.

In the right-angle triangle $\triangle ABC$ (Fig. 3.21), the following formulas are extracted:

$$\tan(90^\circ - \alpha) = \frac{z}{x} \tag{3.2}$$

$$\rightarrow z = x \cdot \tan(90^\circ - \alpha) \tag{3.3}$$

$$\tan(90^\circ - \alpha) = \cot\alpha = \frac{1}{\tan\alpha} \tag{3.4}$$

$$\rightarrow z = x \cdot \cot\alpha \tag{3.5}$$

$$\rightarrow x = \frac{z}{\cot\alpha} \tag{3.6}$$

$$\rightarrow x = z \cdot \tan\alpha \tag{3.7}$$

The same way, in the right-angle triangle $\triangle BCD$ (Fig. 3.21), the base of the triangle is:

$$b - x \tag{3.8}$$

while the following formulas are applied:

$$\tan(90^\circ - \beta) = \frac{z}{b - x} \tag{3.9}$$

$$\stackrel{3.7}{\rightarrow} \cot\beta = \frac{z}{b - z \cdot \tan\alpha} \tag{3.10}$$

$$\rightarrow z = \cot\beta \cdot (b - z \cdot \tan\alpha) \tag{3.11}$$

$$\rightarrow z = \frac{1}{\tan\beta} \cdot (b - z \cdot \tan\alpha) \tag{3.12}$$

$$\rightarrow \tan\alpha + \tan\beta = \frac{b}{z} \tag{3.13}$$

$$\rightarrow z = \frac{b}{\tan\alpha + \tan\beta} \tag{3.14}$$

Finally, in the small right-angle triangle $\triangle DEF$ (Fig. 3.21), the following formula is derived:

$$\rightarrow \tan\beta = \frac{b}{f} \tag{3.15}$$

These 3 equations, namely (3.7), (3.14) and (3.15), are the main formulas used for calculating the distance of each point from the source.

Active range sensors based on ToF and phase-based

This is an example to easily understand ToF laser scanning; this technology can be considered as a single laser range finder. Basically, such a scanning measures a distance by sending out a laser beam to an object and calculating the time it takes the laser beam to bounce back from the surface of the object.

As the laser beam uses light and the speed of light is known, so by measuring the time -the laser beam to bounce back- we can figure out that the distance d is calculated by the multiplication of time t divided by two and multiplying it by c; the constant for the speed of light, as shown in the following Eq. (3.16).

$$\frac{c \cdot t}{2} \tag{3.16}$$

where

d the distance laser-object point
t the time takes the laser beam to bounce back
c the speed light constant, i.e. 299, 792, 458 meters per second

Thus, to compute the X, Y, Z of a point in an object, several elements should be known; the distance to the object, the bearing (horizontal angle from a known line) to the object and the vertical angle to the object.

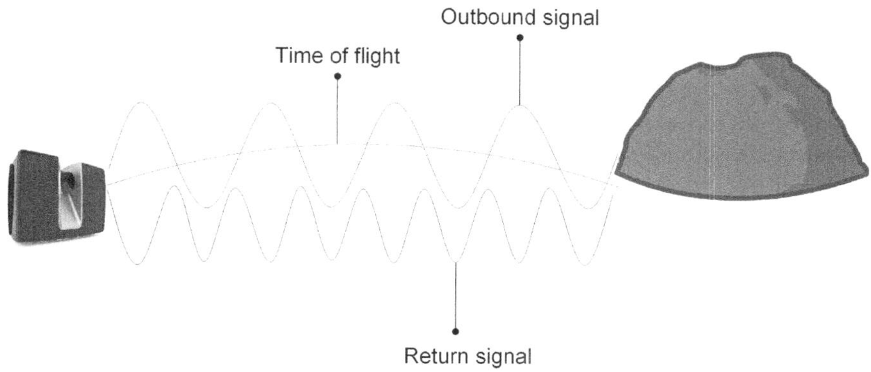

Fig. 3.22 ToF and phase-based laser scanning principle

A ToF laser scanner measures the distance and the angle in horizontal and vertical level for every position it is at. The measurements are captured by the scanner itself while moving in a grid mode in a 360° horizontal plane and roughly 330° in the vertical plane. This is the reason ToF scanning is much slower compared to phase-based scanning. The ToF scanning is functioning for longer ranges, reaching even 300 m. However, the acquisition time is slower, compared to the acquisition time of phase-based scanner.

On the hand, phase-based scanning makes use of a constant beam of laser that is sent out from the laser scanner. Then, the laser scanner measures the phase shift of the returning laser in order to calculate distances. Phase-based laser scanners can capture data at a rate more than a million points per second. However, the range of phase-bases laser scanners is limited to less than 100 m and are mainly suited for building interiors.

The comparison between the ToF and the phase-based laser scanning is illustrated in Fig. 3.22.

3.2.4.4 Mobile Mapping Platforms

Having boomed the recent years, *3D* mobile mapping became one of the most important geospatial technologies. It may combine imaging, positioning, scanning, other measurement technologies and tools and a variety of mobile transportation platforms. *3D* mobile mapping makes possible to measure, record and visualize the captured environments. The blending of the different technologies available and the heterogeneous data fusion aims to alter the way environments are measured, recorded and visualized.

3D mobile mapping is fast, accurate and effective technology, especially for the large environments. The equipment is set on a moving platform, e.g. a car, and this enables significant benefits to the users. Unreachable areas and sizeable areas of the natural and built environment can be mapped in such a manner as to achieve quickly a desired result.

By now, *3D* mobile mapping technology is being used for *3D* surveying and modeling of road networks and railways, for *3D* mapping of urban environments but also underground infrastructures such as tunnels, for *3D* representation of big industrial infrastructures such as power plants, and many more. This technology brings together innovative software, navigation, scanning and visualization platforms in a mobile setting that hands over fast, accurate and concise *3D* representations of the natural and built environment.

3.2.4.5 Unmanned Aerial Vehicles

Geospatial surveys have much altered over several decades. We began with the terrestrial film cameras which were also used in combination with air balloons, kites and homing pigeons. The evolution of sensor and platform technologies has advanced, leading to the deployment of new solutions. These recent innovations include UAVs which are capable of carrying out a complex series of actions automatically. They contributed in completely changing the way data collection is performed, enabling us to accumulate greater detail and providing plentiful insights about the natural and built environment.

In spite of the anticipated progress and growth in UAVs usage, it seems that they will not replace conventional platforms such as vehicles, helicopters, airplanes and satellites. The new technologies are fetching with them some limitations with regard to the number, size and weight of the sensors they can carry as well as some other regulatory restrictions.

As a result, UAVs will continue to offer more and more new capabilities that in a way improve the traditional platforms and provide greater coverage and detail. In addition to that, UAVs are enabling the users to quickly deploy their projects and constantly record and monitor areas where they cannot easily reach (Fig. 3.23).

Fig. 3.23 UAV images covering unreachable areas of a historical building in a traditional settlement

Fig. 3.24 Multirotor and fixed-wing UAVs

UAVs have the ability to deliver high fidelity data with ground sample distance (GSD) of 1 cm and accuracy of under 5 cm. However, accuracy is variable to a high degree as it varies based on the equipment used, the object itself and the software used for data processing. Even though the sensors used for data collection are one of the most crucial factors, many UAVs limit the ability to do multi-sensor flight. The conventional UAVs with an endurance ranging from 20–60 min are able to cover no more than a few square kilometers with a RGB or multispectral camera. The evolution in sensor development is expected to lead to new multi-sensor solutions in a reasonable weight.

In principle, the UAVs are distinguished in multirotor and fixed-wing (Fig. 3.24).

Since the UAVs can be deployed quickly and without having any initialization cost, they rendered a productive, cost-effective and workable solution on a daily basis. Such a solution can be used to cover small areas of the natural and built environment. Besides, the use of UAVs currently faces an important barrier to worthy of attention. The current regulations vary among the different continents and countries: payloads, mandatory UAV registration, line-of-site restrictions, just to mention a few.

The batteries used also play an important role in the usability of UAVs, as their life prospect is affecting the overall performance. The vast majority of most commercial UAVs can fly for only about 20–60 min, regardless of the constant improvements in battery technology. Despite some of the drawbacks, the UAVs have proven the optimum solutions in many use cases, from damage assessment in difficult access points in a huge monument to data collection for big archaeological sites.

In a review paper prepared by Colomina and Molina (2014), the authors discuss the evolution and the state-of-the-art use of UAV in the field of photogrammetry and remote sensing. In this review paper, the authors initially present a concise historic background and analysis of the regulatory status, and then they review the latest developments with respect to the unmanned aircrafts' sensing, navigation, orientation and general data processing, considering the photogrammetric and remote sensing workflows, with emphasis on the nano-micro-mini UAV segment.

Especially for the use of UAV photogrammetry for the documentation of cultural heritage sites, in their work (Federman et al. 2018) elaborate on the way this goal was accomplished through an international case study analysis of two differing sites: Prince of Wales Fort (Churchill, MB, Canada) and Bhaktapur Durbar Square (Kathmandu, Nepal). The selected sites had been documented at an earlier time; however, this was the first time that they used UAV technology for this reason. In their work, they discuss the way in which the images have been acquired, the data processing workflows, and the deliverables generated from the data.

3.2.5 *Tools and Software*

Tools and software constitute a vital infrastructure to process the collected data. The existing tools and software can be distinguished in commercial and open source, while each one uses a different type of algorithm.

When evaluating tools and software used for data and image processing, there are many considerations to examine. One of the biggest decisions is whether to use commercial or open source technology. Before the use of any technology, it is important to savvy the fundamental differences between commercial and open source software. The commercial systems are developed and supported by for-profit companies that typically sell licenses for the use of their software. On the contrary, the open source systems are supervised by devoted communities of developers who provide and contribute modifications to meliorate the product continually. The communities are those who decide on the course of the software based on their needs.

The workflow of carefully selecting the best possible software is to evaluate the available options considering different parameters such as the availability of source

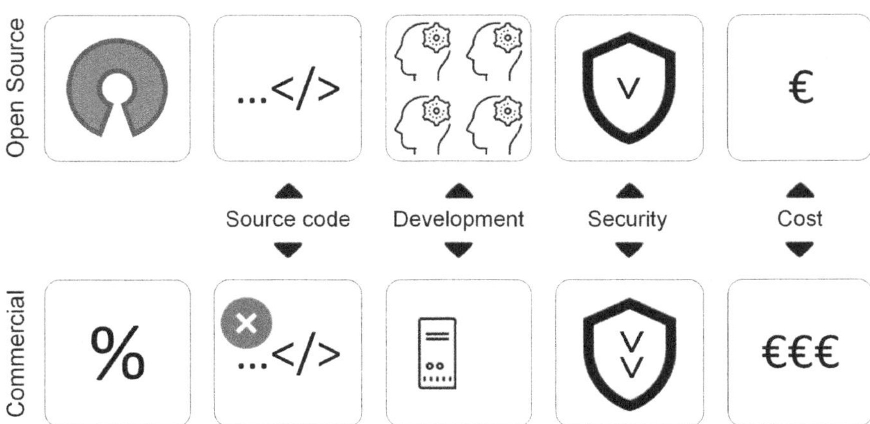

Fig. 3.25 Commercial versus open source software

Table 3.3 3D reality modelling software: commercial and open source

Name	Type
3DF Zephyr	Commercial
COLMAP	Open source
ContextCapture	Commercial
Correlator3D	Commercial
DroneDeploy	Commercial
Imagine photogrammetry	Commercial
iWitness	Commercial
Meshroom	Open source
Metashape	Commercial
MicMac	Open source
MVE	Open source
Photomodeler	Commercial
Pix4D	Commercial
ReCap	Commercial
Regard3D	Open source
OpenMVG	Open source
SURE	Commercial
VisualSFM	Open source

code, the development team and the level of respective support, the security protocols as well as the cost (Fig. 3.25).

The ownership cost is a crucial factor in deciding whether to use open source or commercial software. In most cases, open source software is free or has low cost licensing options. On the other hand, commercial software requires purchasing a license for using the software.

Photogrammetry is an image-based reality modelling technique. By using such a technique, small or large objects can be captured, while it is reasonably priced. A camera is the necessary equipment for image acquisition. However, there is also need for photogrammetry software to process the images and create the *3D* model of the object(s) illustrated in the images.

The software used in photogrammetry workflows can be found in many forms and sizes. Major commercial players are offering commercial solutions that are ideal for many different application areas. Nevertheless, a number of tools are available for free download, based on open source technology. An indicative list is illustrated in Table 3.3.

3.3 Human Threats and Natural Hazards

Nowadays, we are losing the world cultural heritage by human threats and natural disasters. The examples are many where the destruction occurs faster than it can be recorded and documented. Threats and disasters based on humans, such as war and terrorism are sources that are probably the most violent for this loss. Natural disasters, such as floods, are the other source of danger for cultural heritage.

3.3.1 Buddhas of Bamiyan, Afghanistan

In 2001, the civilized world reacted in horror as, part of a campaign to rid Afghanistan of idolatry, the Taliban destroyed the World Heritage Site Buddhas of Bamiyan (UNESCO 2003), as part of the "*Cultural Landscape and Archaeological Remains of the Bamiyan Valley*" (Fig. 3.26). In Afghanistan, we have lost the world's two largest standing Buddhas, one of them 53 m high, the other 35 m, standing over 50 m high above a small town located at the Hindu Kush mountains base of central Afghanistan. The Taliban drilled holes into the two statues' torsos and then placed dynamite in the holes to blow them up.

Fig. 3.26 General view of the Bamiyan Valley. (*Source* UNESCO/CC-BY-SA 3.0)

In the work of the Institute of Geodesy and Photogrammetry, ETH Zurich, Switzerland, (Grün et al. 2002a, b, c, 2004), they explain how *3D* reconstruction of the statues was performed by using archival images. After the destruction in 2001, a consortium was formed, having as its ultimate goal the rebuild of the Great Buddha of Bamiyan, at original shape, size and place. This international initiative was led by the global heritage Internet society New7Wonders (2002), and the Afghanistan Institute & Museum, Bubendorf (Switzerland). The ETHZ group performed the required *3D* reconstruction, which served as the basis for the physical reconstruction of the statue. Their initial target was to use the *3D* model and build the statue at the scale of 1/10. In practice, they produced various outcomes by performing automatic and manual photogrammetric procedures (Fig. 3.27). This scaled statue was displayed in the Afghanistan Museum in Switzerland and was also used for studying the materials and construction techniques to be applicable in the ultimate rebuilding of the statue at full size.

In order to create the *3D* reconstruction model, they used three different types of imagery. This is one of the most typical (international) examples, where amateur, archival and crowdsourcing information was used in such type of cultural heritage documentation and *3D* virtual reconstruction projects.

1. The first imagery dataset consists of four low-resolution images of the big statue, obtained from the Internet as amateur images. These images have different resolutions and sizes, unknown camera constants, and were not captured at the same time.
2. Tourist-type images acquired by Harald Baumgartner who visited the valley of Bamiyan between 1965 and 1969.
3. Three metric images acquired in 1970 by Professor Robert Kostka from the Technical University of Graz in Austria (Kostka 1974).

ICOMOS reported many times on heritage at risk in Afghanistan, particularly on the state of conservation of the giant Buddhas of Bamiyan and the efforts to protect their remains (see Heritage at Risk 2000, pp. 28–42, Heritage at Risk 2001–02, pp. 24–26, Heritage at Risk 2002-03, pp. 16–20, Heritage at Risk 2004-05, pp. 26–31 and Heritage at Risk 2008-10, pp. 16–18) (ICOMOS 2010).

3.3.2 The Plaka Bridge, Greece

The Plaka bridge (Fig. 3.28) was a stone arched bridge in the Arachthos river, in Epirus Region, Northwest of Greece. It is a designated monument under protection by the Greek Ministry of Culture. It was built in 1866 by Kosta Beka, the chief mason, and his coworkers. The arch of the bridge is 40 m wide and 18–20 m high. There were also two small relief arches in each side of 6 m wide. and a top opening of 3.2 m (Leftheris et al. 2006). This bridge was considered the largest one-arched bridge in the Balkans and the third largest in Europe.

(a)

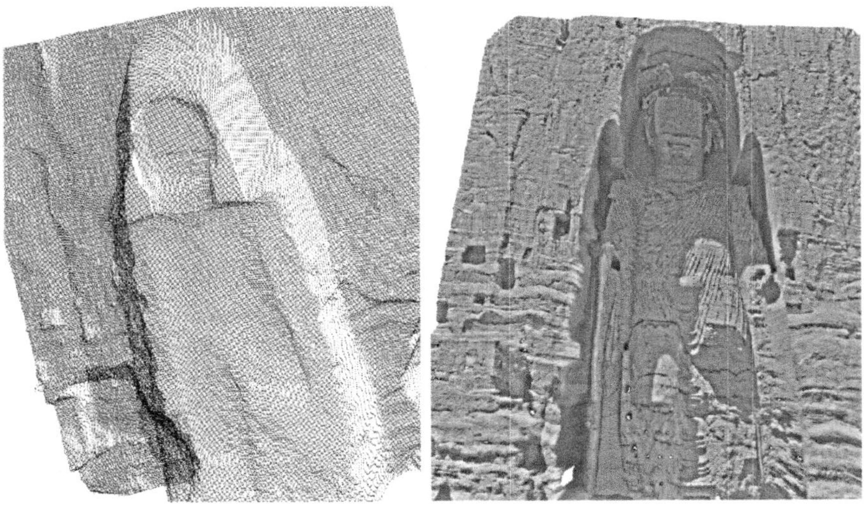

(b)

Fig. 3.27 **a** The cliff of Bamiyan valley with the three Buddha statues and the caves. (*Source* Grün et al. 2002b) **b** 3D point cloud generated with an automatic process on the metric images (left). The detailed folds of the robe are not modeled. The associated photorealistic virtual model (right). (*Source* Grün et al. 2002a)

The bridge collapsed three times: in 1860, in 1863, and in 2015. On the 1st of February of 2015, the central part of the one-arch limestone bridge collapsed. The main reason was the extreme weather conditions in the region and the massive flash flood in the river Arachthos. Next to the main arch that collapsed, there were two smaller ones, the so-called relief arches, which are visible in Fig. 3.28. Most of the parts collapsed, and some of the big ones lie in the river area near the abutments that remain untouched. Two representative images from the bridge condition, just after the collapse, are given in Fig. 3.29, which were retrieved from the Greek new portal http://www.themanews.com.

The significance of the Plaka bridge, the designation from the Greek Ministry of Culture and the uniqueness of the bridge motivated the National Technical University of Athens (NTUA), Greece, to initiate a restoration project. NTUA established an interdisciplinary team pairing the skills and knowledge of different cultural heritage

Fig. 3.28 The Plaka bridge in 2011, before the destruction of 2015. (*Source* Wikipedia/CC-BY-SA 2.0)

experts to investigate all the scientific and technical parameters for the potential reconstruction of the bridge. This team had the mandate to carry out a systematic work on the constructional and structural analysis of the bridge, which was based on the architectural plans. For this reason, a geometric, architectural and structural data acquisition, process and analysis took place in two stages, before and after the collapse of the bridge (Kouimtzoglou et al. 2017).

The study was challenging for the NTUA team for many reasons, but especially for the multiple and heterogeneous data sources. In 1984, 1995, and 2005, three different studies and surveying campaigns were held. In those three cases, conventional (topographical) equipment was used, while the results and outcomes should have been evaluated and paired with the new acquired data and findings. However, the new survey was only able to touch the remaining parts of the bridge after the collapse, as illustrated in Fig. 3.29. In addition to that, the designs from the previous campaigns were considered incomplete due to the lack of critical details such as the stonework, the metal and wood reinforcement systems, etc. The older survey campaigns were only focused on the precise outline of the bridge structure, without focusing on details which are necessary for an architectural restoration project (Kouimtzoglou et al. 2017).

The interdisciplinary NTUA team was formed by the 5 different Schools, namely:

- Architecture

(a)

(b)

Fig. 3.29 **a** The Plaka bridge just after the collapse where river Arachthos overflew. **b** The bronze sign of the Greek Ministry of Culture illustrating that the bridge is a designated monument. (*Source* (Protothema 2015))

- Civil Engineering
- Rural and Surveying Engineering
- Chemical Engineering
- Mining and Metallurgical Engineering.

Just after the Plaka bridge collapse, the team initiated a crowdsourcing campaign to collect as many images as possible from tourists and amateur photographers who visited the Plaka bridge before it collapsed in 2015. In order to provide a suitable framework to host the images, they developed a website (in Greek), available at http://gefyriplakas.ntua.gr. The first month of the website operation, they managed to collect from more than 130 contributors. In addition to that, nearly 200 images and 15 videos were collected through other means, such as post mail delivery, personal contacts, etc. (Stathopoulou et al. 2015).

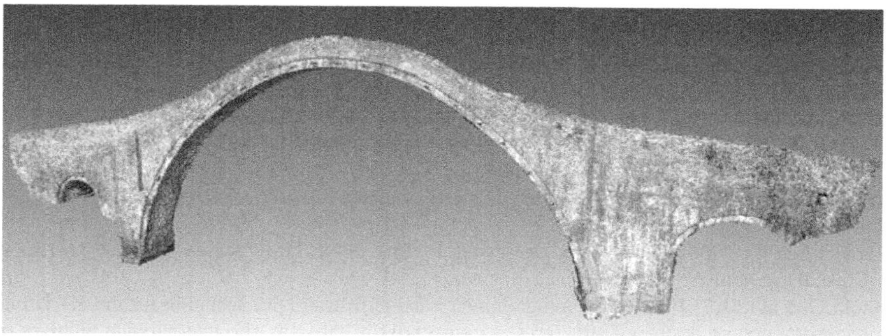

(a)

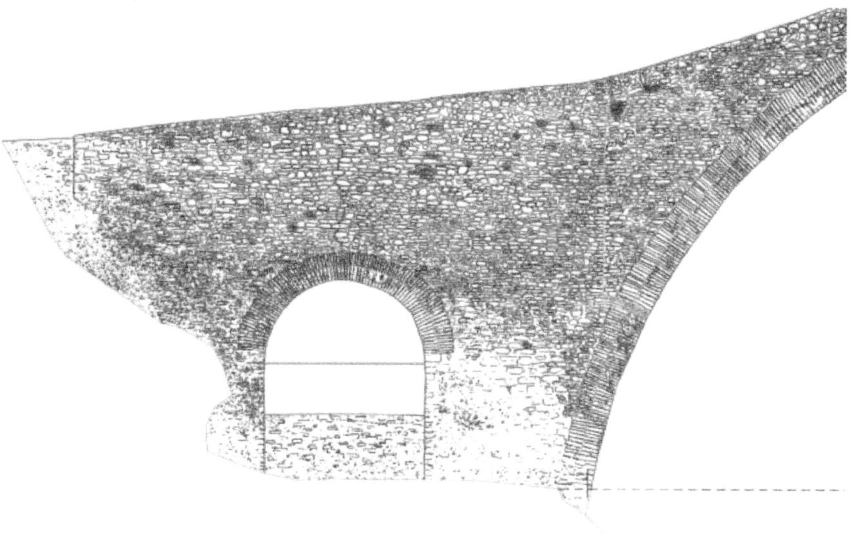

(b)

Fig. 3.30 **a** The 3D point cloud of the Plaka bridge by using the images from the crowdsourcing campaign. **b** A pathology elevation plan of the South façade before the collapse. (*Source* (Stathopoulou et al. 2015))

The NTUA team managed to produce a series of outcomes ranging from detailed topographical and architectural designs and structural analysis plans to point cloud and orthoimages of different parts of the bridge. A technical report was also part of this study. Two of the many outcomes are provided in Fig. 3.30: a *3D* point cloud which was generated by the photogrammetric processing of the crowdsourced images and a pathology elevation plan for one specific part of the bridge.

References

AKDN (2019) Aga Khan foundation. Accessed 13 Mar 2019. http://www.akdn.org

Aparac-Jelušić T (2017) Digital libraries for cultural heritage. development, outcomes, and challenges from European perspectives. Morgan & Claypool Publishers, San Rafael, 203 pp. ISBN: 9781681730837. https://doi.org/10.2200/S00775ED1V01Y201704ICR058

Arches (2019) Arches: heritage inventory and management system. Accessed 14 Mar 2019. http://www.archesproject.org

Berners-Lee T (2006) Linked data. Accessed 13 Mar 2018. http://www.w3.org/DesignIssues/LinkedData.html

Bitelli G et al (2017) Historical photogrammetry and terrestrial laser scanning for the 3D virtual reconstruction of destroyed structures: a case study in Italy. In: The international archives of the photogrammetry, remote sensing and spatial information sciences, vol XLII.5/W1. Florence, Italy, pp 113–119

Boehler W, Heinz G (1999) Documentation, surveying, photogrammetry. In: Proceedings 17th international symposium CIPA 1999. Olinda, Brazil

Boehler W, Marbs A (2004) 3D scanning and photogrammetry for heritage recording: a comparison. In: 12th International conference on geoinformatics - geospatial information research: bridging the pacific and atlantic. Gavle, Sweden, pp 291–297

Canciani M, Saccone M (2011) The use of 3D models in integrated survey: the church of st. thomas of villanova in castel gandolfo. In: The international archives of the photogrammetry, remote sensing and spatial information sciences, vol XXXVIII.5/W16. Trento, Italy, pp 591–597

Colomina I, Molina P (2014) Unmanned aerial systems for photogrammetry and remote sensing: a review. ISPRS J Photogramm Remote Sens 92:79–97. https://doi.org/10.1016/j.isprsjprs.2014.02.013

CreativeCommons (2018) Creative commons. Accessed 28 Feb 2018. https://creativecommons.org

EC (2018) Enforcement of intellectual property rights. Accessed 19 Mar 2018. https://ec.europa.eu/growth/industry/intellectual-property/enforcement_en

Europeana (2018) Europeana. Accessed 26 Feb 2018. https://pro.europeana.eu/our-mission/history

Faro (2018) FARO scanner freestyle3D X. Accessed 26 Mar 2018. https://www.faro.com/products/construction-bim-cim/faro-scanner-freestyle3d-x/

Federman A et al (2018) Unmanned aerial vehicles (UAV) photogrammetry in the conservation of historic places: carleton iimmersive media studio case studies. Drones 2(2):774–781. https://doi.org/10.3390/drones2020018

Getty JP (2018) The getty. Accessed 18 Mar 2018. http://www.getty.edu

Grün A, Remondino F, Zhang L (2004) Photogrammetric reconstruction of the Great Buddha of Bamiyan, Afghanistan. Photogramm Rec 19(107):177–199

Grün A, Remondino F, Zhang L (2002a) Reconstruction of the Great Buddha of Bamiyan, Afghanistan. In: International archives of photogrammetry and remote sensing, vol 34.2. Corfu, Greece, pp 363–368

Grün A, Remondino F, Zhang L (2002b) The reconstruction of the Great Buddha of Bamiyan, Afghanistan. In: ICOMOS international symposium. Madrid, Spain

Grün A, Remondino F, Zhang L (2002c). Image-based reconstruction and modelling of the great buddha statue in Bamiyan, Afghanistan. In: International archives of photogrammetry and remote sensing, vol XXXIV.5/W12. Torino, Italy, pp 173–175

Grussenmeyer P et al (2016) Basics of rangebased modelling techniques in cultural heritage recording. In: Stylianidis E, Remondino F (eds) 3D recording, documentation and management of cultural heritage. Whittles Publishing, Dunbeath, pp 305–368

Grussenmeyer P, Jasmine J (2003) The restoration of beaufort castle (South-Lebanon): a 3D restitution according to historical documentation. In: Proceedings XIXth CIPA international symposium 2003. Antalya, Turkey

Hanke K, Moser M, Rampold R (2015) Historic photos and TLS data fusion for the 3D reconstruction of a monastery altar ensemble. In: The international archives of the photogrammetry, remote sensing and spatial information sciences, vol XL.5/W7. Taipei, Taiwan, pp 201–206
Hexagon (2018) AICON SmartScan. Accessed 25 Mar 2018. http://www.hexagonmi.com/products/white-light-scanner-systems/aicon-smartscan
Howe J (2006) Crowdsourcing: a definition. Accessed 25 Feb 2018. http://crowdsourcing.typepad.com/cs/2006/06/crowdsourcing_a.html
Howe J (2006b) The rise of crowdsourcing. In: Wired (ed). Accessed 25 Feb 2018. https://www.wired.com/2006/06/crowds/
ICA (2018) International cartographic association. Accessed 20 Mar 2018. https://icaci.org/mission/
ICOM (2018a) International council of museums. Accessed 18 Mar 2018. http://icom.museum
ICOM (2018b) ICOM code of ethics for museums. Accessed 19 Mar 2018. http://icom.museum/the-vision/code-of-ethics/
ICOMOS (2010) Heritage at risk. ICOMOSWorld report 2008–962010 on monuments and sites in danger. Accessed 14 Mar 2018
JMK (2019) J.M. Kaplan fund. Accessed 13 Mar 2019. http://www.jmkfund.org
Konica-Minolta (2018) Non-contact 3D digitizer vivid 910/VI-910. Accessed 25 Mar 2018. https://www.konicaminolta.com/instruments/download/instruction_manual/3d/pdf/vivid-910_vi-910_instruction_eng.pdf
Kostka R (1974) Die stereophotogrammetrische Aufnahme des Grossen Buddha in Bamiyan. Afghan J 3(1):65–74
Kouimtzoglou T et al (2017) Image-based 3D reconstruction data as an analysis and documentation tool for architects: the case of Plaka bridge in Greece. In: The international archives of the photogrammetry, remote sensing and spatial information sciences, vol. XLII.2/W3. Nafplio, Greece, pp 391–397
Leftheris BP et al (2006). In: Computational mechanics for heritage structures. WIT Press, Southampton, 288 pp. ISBN: 978-1-84564-034-7
Leica (2018) Leica nova MS60. Accessed 21 Mar 2018. https://leica-geosystems.com/products/total-stations/multistation/leica-nova-ms60
LinkedData (2018) Linked data -connect distributed data across the web. Accessed 11 Mar 2018. http://linkeddata.org/
Louvre (2018) Louvre museum. Accessed 19 Mar 2018. http://www.louvre.fr/en
Marden J et al (2013) Linked open data for cultural heritage. In: SIGDOC'13 proceedings of the 31st ACM international conference on design of communication. Greenville, North Carolina, USA, pp 107–112. ISBN: 978-1-4503-2131-0. https://doi.org/10.1145/2507065.2507103
MET (2019) The metropolitan museum of art: how to read (and display) architectural drawings. Accessed 14 Mar 2019. http://www.metmuseum.org
Myers D et al (2012) Arches: an open source GIS for the inventory and management of immovable cultural heritage. In: Ioannides M et al (ed) Progress in cultural heritage preservation, vol. 7616. Springer, Berlin, pp 817–96824
New7Wonders (2002) Bamiyan buddha project. Accessed 14 Mar 2018. https://about.new7wonders.com/new7wonders-bamiyan-buddha-project/
NextEngine (2018) NextEngine 3D scanner ulta HD. Accessed 25 Mar 2018. http://www.nextengine.com
NYPL (2018) New York public library. Accessed 27 Feb 2018. https://www.nypl.org
Pentax (2018) Pentax series R-400VDN. Accessed 21 Mar 2018. http://pentaxsurveying.eu.com/en/index.php/products/total-stations/r-400vdn/
Protothema (2015) Waters, wind cause collapse of historic Plaka bridge in Arta. Accessed 15 Mar 2018. http://en.protothema.gr/historic-plaka-bridge-in-arta-damaged-by-extreme-weather-conditions-photos-video/
Ridge M (2014) Crowdsourcing our cultural heritage: introduction. In: Ridge M (ed) Crowdsourcing our cultural heritage. Ashgate, Surrey, pp 1–9613. ISBN: 9781472410221

SAA (2018) Society of American archivists. Accessed 26 Feb 2018. https://www2.archivists.org/standards/DACS/statement_of_principles

Sansoni G, Trebeschi M, Docchio F (2009) State-of-the-art and applications of 3D imaging sensors in industry, cultural heritage, medicine, and criminal investigation. Sensors 9(1):568–601. https://doi.org/10.3390/s90100568

Stathopoulou E et al (2015) Crowdsourcing lost cultural heritage. In: ISPRS annals of the photogrammetry, remote sensing and spatial information sciences, vol. II.5/W3. Taipei, Taiwan, pp 295–300

Themistocleous K (2017) Model reconstruction for 3D vizualization of cultural heritage sites using open data from social media: the case study of soli, Cyprus. J Archaeol Sci: Rep 14:774–781. https://doi.org/10.1016/j.jasrep.2016.08.045

Topcon (2018) Topcon IS-3 imaging robotic total station. Accessed 21 Mar 2018. http://eu-dev-uk.topconpositioning.com/total-stations/robotic-total-stations/3-series

Trimble (2018) Trimble SX10 scanning total station. Accessed 21 Mar 2018. https://geospatial.trimble.com/products-and-solutions/sx10

UNESCO (2003) Cultural landscape and archaeological remains of the bamiyan valley. Accessed 14 Mar 2018. http://whc.unesco.org/en/list/208

UNESCO (2005) Convention on the protection and promotion of the diversity of cultural expressions. Accessed 01 Mar 2018. http://portal.unesco.org/en/ev.php-URL_ID=31038&URL_DO=DO_TOPIC&URL_SECTION=201.html

Vincent M, Coughenour C (2018) Rekrei. Accessed 26 Feb 2018. https://projectmosul.org/

Webfoundation (2018) World wide web foundation - Sir Tim Berners-Lee. Accessed 11 Mar 2018. https://webfoundation.org/about/sir-tim-berners-lee/

Wiedemann A, Hemmleb M, Albertz J (2000) Reconstruction of historical buildings based on images from the meydenbauer archives. In: International archives of photogrammetry and remote sensing, vol 33.B5/2, pp 887–893

Wikipedia (2018) Wikipedia. Accessed 26 Feb 2018, USA. https://en.wikipedia.org/wiki/Main_Page

Williams C (2006) Managing archives foundations, principles and practice, 1st edn. Chandos Publishing, Oxford, 247 pp. ISBN: 1 84344 112 3

WIPO (1883a) Paris convention for the protection of industrial property. Accessed 01 Mar 2018. http://www.wipo.int/treaties/en/ip/paris/

WIPO (1886b) Berne convention for the protection of literary and artistic works. Accessed 01 Mar 2018. http://www.wipo.int/treaties/en/ip/berne/

WIPO (1961c) Rome convention for the protection of performers, producers of phonograms and broadcasting organizations. Accessed 01 Mar 2018. http://www.wipo.int/treaties/en/ip/rome/

WIPO (2018d) World intellectual property organization. Accessed 27 Feb 2018. http://www.wipo.int

Chapter 4
Historic Buildings

Abstract A vigilant look at the parameters defining a building as historic is discussed in this chapter. The characterization of a building as historic has a direct effect on the surveying and documentation of the building. Several factors are described in this chapter to designate the architectural, historic and cultural interest of such a building. Factors such as age, architectural character, history, and uniqueness are discussed within this chapter. Building location and sense are of great importance, potentially for the inhabitants, a city or even a nation. Setting of a building is critical as well, discussed also in the ICOMOS Bura Charter. Building identity as well as materials and construction itself are also deliberated in this chapter. Building pathology is a holistic approach to understanding buildings while the historic building information systems are aiming to support specific tasks, management and decision-making.

Before beginning the technical part of this study, a careful look at the parameters rendering a building *historic* is necessary. The definition of such a framework is supporting the consideration of what effect this '*historicalness*' will have not only in inspection, analysis, assessment and reporting, but mainly in the surveying and documentation of the building. The emphasis is given in the surveying dimension, the so-called geometric documentation, since this is the main concern of this book.

4.1 Architectural, Historic and Cultural Interest

The factors described in the following are affecting the architectural, historic and cultural interest of the building. They are illustrated in alphabetical order and not in terms of significance.

E. Stylianidis, *Photogrammetric Survey for the Recording and Documentation of Historic Buildings*, Springer Tracts in Civil Engineering,
https://doi.org/10.1007/978-3-030-47310-5_4

4.1.1 Age

Age is a factor that makes a building *historic*. It is considered along with architectural, historical and cultural interest with important people and events. Imagine the place (building) where a great philosopher, poet, composer or an outstanding President was born or lived. In addition, the place where a momentous historical event took place, for example, the starting point of a revolution.

Normally, in order to nominate for designation, a building must be at least few decades old. The ageing limits to be qualified as historic vary from country to country. In some countries, they have a double system, e.g. a national register and a local designation. They are two entirely separate processes.

4.1.2 Architectural Character

The Technical Preservation Services, National Park Service, U.S. Department of Interior, published *Preservation Briefs* (NPSa 2018). In fact, it is a guidance on preserving, rehabilitating, and restoring historic buildings. The briefs are 50 useful steps that should be followed in historic building preservation.

One of these 50 briefs, no.17, is titled "*Architectural character—Identifying the visual aspects of historic buildings as an aid to preserving their character*" (NPSb 1988). The editors of the publication are underlining that "*every old building is unique, with its own identity and its own distinctive character. Character refers to all those visual aspects and physical features that comprise the appearance of every historic building. Character-defining elements include the overall shape of the building, its materials, craftsmanship, decorative details, interior spaces and features, as well as the various aspects of its site and environment*".

4.1.3 History

Time and place are highly related to a historic building by providing meaning and purpose. History elucidates and illustrates events of the past. This is why we consider buildings as an integral part of history: an information stratigraphy that should be unlocked. Inspection, recording and documentation define the necessary path to interpret, comprehend and share this architectural, historic and cultural knowledge.

4.1.4 Uniqueness

Uniqueness of the building type, a peculiar structural system, its artistry, some exceptional decoration (e.g. interior or exterior paintings), or even the relationship of the

Fig. 4.1 Snapshot from a building facade designed by Antoni Gaudi (*Source* Pixabay/CC0 Creative Commons)

building to its setting are critical parameters for characterizing a building as 'unique'. A representative example of uniqueness in architecture is the masterpieces of Antoni Gaudi in Spain (Fig. 4.1). Gaudi's work is not only unique for Barcelona compared to other Spanish cities, it is also globally unique. Besides, this is one reason why Barcelona is visited by millions of tourists and thousands of professionals (e.g. architects, artists, etc.) from all around the world every year.

4.2 Building Location and Sense

Some buildings pose a strong sense for their particular position and place. This is of great significance and value for both the inhabitants and the town or even the whole country. It is also important because it has the power to be understandable and enjoyable for the visitors and the tourists. The features combined to form this condition are various and are mainly related to the location sense of the building.

- Buildings that continue to exist bringing strong images, memories, or feelings to mind of an event. Creations, destructions, and strong historical or political events are of great extent on the list of examples, notably if physical evidences persist

Fig. 4.2 Exterior and interior view from the Essex County Jail, Newark, New Jersey, US

in an activity to reflect the modern state of affairs and the building has not been totally collapsed.

- Buildings where operation or belief are the most important, powerful and influential elements in granting expression to the configuration, nature and location of the building. A characteristic of its kind are buildings like churches, old industrial installations and jails, etc.
- Buildings that are dominated or influenced by a strong and distinctive attribute, which can be either natural or human-made, that imposes its character on the location and surroundings.
- Buildings that have been fashioned and used by the dwellers for a period of time in a manner conforming with local characteristics, materials and beliefs from generation to generation. Possibly, their characteristics may not be particularly interesting or surprising, but connected to the level to which the location provides a sense of continuity and identity for its dwellers.

Essex County Jail (Fig. 4.2) is located in Essex County, New Jersey, United States, and it is Essex County's oldest public building. The original building was built in 1837, and in 1890 it was expanded with several additions increasing the number of prison cells up to 300. In 1970, the jail was abandoned when a new county jail was built. On September 3, 1991, the building was added to the National Register of Historic Places.

In spring 2018, when I was on my sabbatical leave at Columbia University, a team of graduate students from the Graduate School of Architecture, Planning and

Preservation (GSAPP), Columbia University, performed the documentation and researched the architectural and social history of this structure. The images on Fig. 4.2 are from that period.

4.2.1 Setting

According to the Burra Charter (ICOMOS 2013), "*Setting means the immediate and extended environment of a place that is part of or contributes to its cultural significance and distinctive character*" (Article 1.12). In fact, it is an area all around a particular place whose limits are mainly decided by sensory standards, i.e. by using three human senses: vision, hearing, and olfaction. Setting may contain as part of a whole, structures (constructions), spaces, land, water and sky; the visual setting containing the ability to see the place, and other sensory particular features of the setting related to human hearing and olfaction. In addition to these, it may also comprise historic and modern relationships with tangible or intangible aspects of the place.

Setting is vulnerable to an increasing extent, especially nowadays where use and income from the operation of the available space is more than obvious. In any case, a setting should be preserved with relevant respect. For this reason, a clear understanding of the nature and significance of the place is required. Also important is that a setting should be reinforced rather than maintained.

The important topic of the building setting and the protection of the setting emerged from the Venice Charter (Sect. 2.3.2) (Annex A) as an imperative for protection. The building suggests a particular historic relationship. Apart from this, the environment in which it is located does the same. Any alteration in the setting can possibly result in a distorted or different explanation of the meaning and the clearly visible significance of the building.

When considering a potential alteration to a setting, the decision-making process should be equipped with the appropriate tools and careful thoughts. Any change or modification could be intolerable for the place and thus not acceptable. Presumably, any change designed is not be pursuant with the features and functions that made the place significant.

4.3 Building Identity

A building declares what it is and what it stands for. This is building identity. It has a function, and it represents a notion of how it looks and works. The relationship between these can make buildings very special, especially when dealing with historic buildings. It is also the case where people, societies and states stimulate existential inquiries, as, for example, when the building has an ethnic and symbolic meaning.

Typical buildings such governmental offices focus ideally on customers, i.e. the citizens who every day require public services and the efficiency of the provided services. Another example is the hospitals where the interest is for the patients to ease their presence in this place.

The identity of a historic building is established by several features such as the form, size, scale and decorative features in the building.

4.4 Building Materials and Construction

Building materials have always had an extremely important role in the construction of historic buildings but also in the development and recognition of architectural styles and trends. They are used to make up building elements and structures using different processes and techniques that have evolved over many centuries to create a discreet architectural style (Fig. 4.3).

A historic building has its origin in a cultural heritage period which was constructed using the technology of that period, the dexterity of its builders, and also the specific materials deployed as a means of accomplishing the construction of the building.

The careful choice and use of building materials in a particular era within a specific geographic area or for a clearly defined (historic) reason reflects the preferences of the dwellers of this area. It also indicates an awareness of financial and pragmatic necessities of the epoch and the human ability to create by using specific building materials. The term traditional material is used to show the ability of perceiving the meaning of raw materials that are used to create the building materials in continuous practice for a certain purpose, i.e. building construction. Based on the characteristics of the place and the natural sources, such materials are generally of natural origin and porous. Frequently, they have textures and colours that complement the landscape of the place and are eventually sensitive to decay.

The gradual development of building materials and construction techniques is an extremely interesting subject. Before the advent of manufactured building materials, almost all buildings were constructed from traditional materials: marble, wood, animal skins, stone, mud, etc. However, the aim of this book is not to analyze all of the traditional materials.

Especially for the buildings, five big different categories of materials could be classified (in alphabetical order):

1. Concrete and cast stone
2. Finishes
3. Masonry
4. Metals
5. Natural and synthetic polymers

The development of these five categories is illustrated in Fig. 4.4, while a short description for each material is provided in the following.

Fig. 4.3 Stone and half-timbered dwellings (*Source* Pixabay/CC0 Creative Commons)

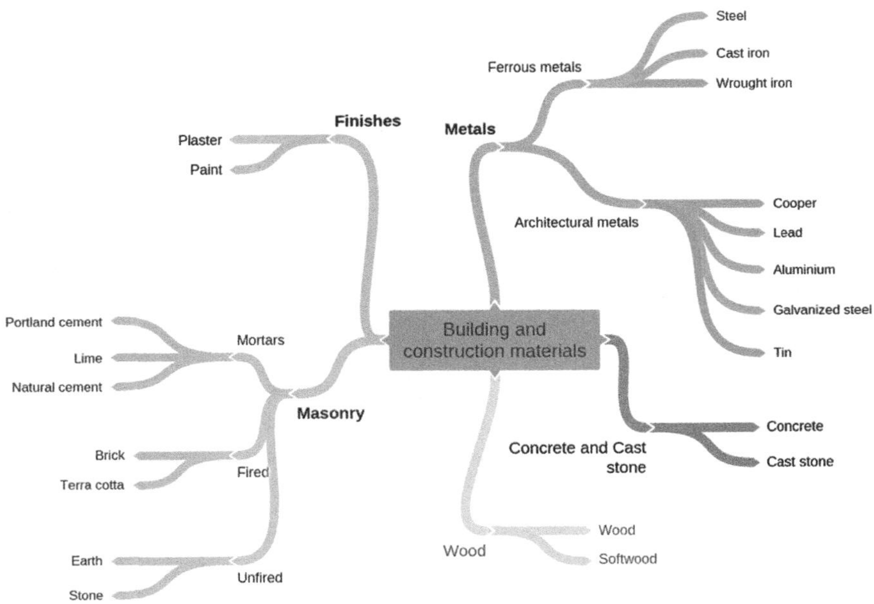

Fig. 4.4 An overview of building materials and construction

- **Metals**: Historically, objects which were made from iron or steel were created for household, religious, artistic, technical/construction but also military purposes.
  - **Ferrous metals**

    Steel: It is an alloy of iron, carbon, and supplementary materials, such as chromium, manganese, nickel, and molybdenum. Carbon is the main element for making steel hard, while as the carbon content increases, the hardness of the steel increases as well. However, ductility and weldability decreases, which means that when subjected to great force, it will not suddenly crack, but slowly bend out of shape. The main problem with this type of steel (carbon) is that it is more sensitive to corrosion than the galvanized aluminum, or stainless steel. Most building constructions are made with a type of steel called mild steel. This type of material is extremely strong. This immense strength of steel is excellently an advantage to buildings. The flexibility of steel framing in a building is another significant distinct attribute as it may bend without cracking. This is extremely important and a great advantage to the building, as it is able to bend when force is exerted on the building (e.g. by an earthquake).

    Cast iron: Cast iron is a group of iron-carbon alloys where the carbon content is increased to that typical range of $2 - 4\%C$ (Smallman and Ngan 2007). It has a relatively low melting temperature. According to Wagner (1996), the development of cast iron metallurgy is dated back many centuries ago in ancient China. The earliest cast iron artifact across the world originated from the tomb of Luhe County, Jiangsu, an eastern-central coastal province in China, dated to the early 5th century B.C. By the early 3rd century B.C., probably earlier, cast iron was extensively used for manufacturing tools and agricultural implements. In the West, cast iron did not become an important material before as late as the 14th century A.D. In architecture and the construction industry, cast-iron architecture is an architectural form which became famous during the Industrial Revolution (1760–1840) when the material was comparatively inexpensive and modern steel had not been developed yet. It was successfully used for certain structures such as buildings and bridges.

- **Architectural metals**

  Copper: The chemical symbol for copper is *Cu*. It is originated from the Latin word *cuprum*, meaning from Cyprus, which is the place where the Romans acquired much of their copper. Copper has been an absolutely necessary material since prehistoric times. It was the first metal used by human beings in any quantity, and in spite of the fact that iron was the basic metal of every Western civilisation from Rome onwards, it was the copper metals which were used when an amalgamation of strength and resistance was considered essential. Copper has the ability to withstand the effect of corrosion, and for this reason remained a functional and decorative material to the present day, in architecture and building construction (cathedrals, castles, roofs, gutters, domes, etc.) (Cu 2018).

  Lead: It is aFVn element that occurs throughout the environment in a natural manner. Lead is an easy to mold, tractable, heavy metal, with a density that goes beyond that of most commonly used materials. Lead may have been used in historic buildings, in the paint, roof and pipework. A number of material properties such as high density, low melting point, ductility, resistance to oxidation, etc., rendered lead as a practical material in construction for hundreds of years. However, in the late 19th century, lead was observed as toxic, and its use has been less diffuse since then.

  Aluminium: It is a light, durable and functional material. Nowadays, it is considered one of the key engineering materials, and it can be found in houses, in automobiles, in trains and aeroplanes, in mobile phones and computers, etc., but a mere 200 years ago, very little was known about this metal. Compared to other metals, aluminum was discovered in recent years. It is considered as the third most abundant element on Earth, after oxygen and silicon. Pure aluminum is not used for architectural applications because its mechanical strength is very low. For this reason, aluminium is mixed with other metals such as silicon and copper to increase its strength, while the alloys can be either wrought or cast. When exposed to the atmosphere, pure aluminium is very sensitive to corrosion, and finishing of aluminum is necessary to impart corrosion resistance (Flandro and Thomas-Haney 2015).

Galvanized steel: It is a carbon steel that has been provided with a layer of zinc. The most widespread method of zinc coating is the hot-dip process. A hot-dipped coated metal provides better resistance to corrosive environments. The spangle finish on the surface makes the material visually attractive, but also makes it more solid and durable from corrosion. Typical uses for galvanized steel in building and construction are wood connectors, rain gutters, etc.

Tin: Tin is a soft, malleable, ductile metal. From the Bronze Age and onwards, tin has been a vital metal in the creation of tin bronzes. The use of tin is dated very far back to around 3000 B.C in the Middle East and the Balkans, and it was a significant part of the ancient cultures. Bronze was the first tin alloy used on a large scale which was made of tin and copper and used in buildings.

- **Masonry**: It is the art of building and construction in stone, clay, brick, or concrete. Many times, the construction of poured concrete, reinforced or not, is also assumed masonry. Broadly, masonry is used to form the walls and other solid elements of buildings and structures.

  - **Fired**

    Brick: *Brick* is the dominant wall and foundation material in many Western cities, especially in USA (New York, Washington, etc.). It can be found in many different types, concerning color, size, shape, and textures. There are several forms such as pressed, common, Roman, etc. Usually, bricks were manufactured using either steel or wood molds, which form the shape, the size and the texture of the brick. The clay, i.e. the raw material, usually is originating from the broader area. The color and the hardness of the brick are influenced by the type of the clay and the temperature of the kiln. Glazed brick is manufactured by adding glaze to the finished brick and then by re-firing once again. It can be also found in a wide range of colors.

    Terra-cotta: *Terra-cotta* is a type of pottery made of clay fired to a porous state, and it can be unglazed or glazed ceramic. Under normal conditions, terra-cotta is employed for sculptures made in earthenware; however, it is used in many other occasions in building construction (bricks, roofing tiles, etc.). Glazed architectural terra-cotta has significant influence in the process of developing momentous architectural forms and characteristic modes of expression. Glazed architectural terra-cotta and its unglazed particular form in exterior building surfaces was used in Asia for several centuries before becoming a trend in the West in the 19th century. Architectural terra-cotta also concerns the ceramic decorations in temples and other structures in the classical architecture of Europe, as well as in the ancient near East. Glazed architectural terra-cotta has many material features similar to brick and stone. It also has many material properties substantially different from those traditional materials in masonry. Terra-cotta may be natural in color, i.e. brown-red, and this is the reason of its name, or glazed in a large range of colors.

- **Mortars**

  Lime: *Lime* mortar is made of lime, sand and water. It is a traditional material in mortars and renders, with its use dating back many centuries. As a building construction material, it has been known since the construction of Jericho in 7000 B.C. (Sickels-Taves and Allsopp 2005). With reference to past events, lime was used in different forms, while periodically, lime mortars were reinforced with a variety of additional substances. Lime continues to exist as the key binding substance for mortar; however, the development of Portland cement in the 19th century gradually took over the place or added an extra element to lime, and much of the ancient trade's knowledge has been lost. Many times, this occurred to the detriment of masonry construction and conservation (Jordan 2005). Nowadays, lime mortar is mainly used in the conservation of buildings that were originally built with lime mortar, but it may also be used as an alternative to common Portland cement.
  Natural cement: *Cement*, in general, undoubtedly has remodeled the human practical contact with building materials, and it could be discussed, that is, at least to some extent, responsible for the introduction in the modern, industrialized world. But what renders a cement "natural"? The natural determination points out that the raw material, a type of limestone acquainted as clayey marl, is mined and burnt with no further processes and additions. On the contrary, Portland and other "artificial" cements are produced from a synthetic mixture of pure limestone, silicates and clays. Natural cements first came into extensive manufacture and usage in Europe and soon spread to the US. Many famous national monuments and infrastructure projects, like the Washington Monument and the Brooklyn Bridge (the towers), used natural cement.
  Portland cement: *Portland cement* is the most ordinary cement type concerning the use in building construction across the world. It is the primary ingredient of concrete, stucco, and other building materials. Portland cement was developed in England, and it was patented in 1824 by Joseph Aspdin. He named it so because of the resemblance in color to the stone from the Isle of Portland, in the South England. More or less, the process remains in essence the same. Limestone and clay are mixed, pulverized, and burned to form a clinker, which is then ground to a powder to make cement. The most common of this type, which is called ordinary Portland cement, is grey in color, but likewise white Portland cement is available. In the US, it came into common use in the early 1900s and by the 1940s was pretty universal.

- **Unfired**

  Adobe: Ancient humans used *adobe* as one of the first materials in order to create buildings. This is dated back centuries ago B.C. It is considered as a low-cost and environmentally friendly building method. In fact, it is nothing more than touchable bricks that are made of sun-dried mud. Adobe bricks are formed by a particular process, if water is mixed with sand and clay.

Straw or grass is added under normal conditions. This helps the mud shrink into uniform brick shapes as it becomes dry. The mud mixture is placed into wooden forms and flatten by hand. Then, the bricks are removed from the forms and placed on a surface in the sun in order to dry. After that, the bricks are set for at least four weeks for air-drying. A building made by adobe helps keep buildings cool in a natural manner during summer and warm in winter. This contributes to reducing the requirements for air conditioning and heat.
Earth: Typically, *earth mortars* are made of local physical subsoil, where clay minerals act as the binding substance to sand and silt particles. The top layer of soil contains organic physical substances, are unstable and are never used for this reason. If an earth mortar contains a good range of particle sizes and not simply fine particles, it is expected to have good working qualities and prove durable. On the contrary, if it is badly classified, it will always be susceptible to physical disaster and to progressive decay.
Stone: History itself has been marked on the surface of a *stone* (Fig. 4.5) from prehistoric monuments to modern buildings. Different types of stone have been used to build, to clad, and to decorate a building. Indicative examples are limestones and sandstones, marbles and granites. Looking around the world, the buildings that are symbols of a country, a district, or a city are mainly built of stone. The Acropolis in Athens is the symbol of the city, the same counts for the Coliseum in Rome, but also Machu Picchu in Peru. There are thousands of examples across the world. Stone can be used in many different forms; from traditional mortared stone walls, to traditional dry-stack stone walls, to framed-one side stone walls, to stone houses, and many more. Primarily, the historical use of stone related to the closeness of the material resources to the places where these were required and the transport easiness (Winter et al. 2014). No other building material rivals the stone for a combination of quality and beauty, stableness, and enduring long-lasting admiration.

- **Wood**
  - **Hardwood**: *Hardwood* is one of the most prevalent types of wood used in fabricating. All structural elements from oak to mahogany are constructed with hardwood. Other types of hardwood trees include beech and birch. One reason for naming hardwoods is because they grow a lot more slowly than other types of trees, which results in denser trunk, bark and branches. Because of the enhanced density of the tree, the wood is heavier, and that is why it is typically used in furniture.
  - **Softwood**: *Softwood* is a type of wood that is cut from trees. A typical example is coniferous trees. Softwood trees, such as pine, cedar, and many more, continue to have their leaves throughout the entire year. It provides the vast majority of all timber, and under normal conditions is supplied in long, rectangular forms such as, for instance, planks. Ordinarily, it is used in the construction industry, in the roof and inner walls of structures, as well as in other building components, such as doors.

Fig. 4.5 Stone masonry (*Source* Pixabay/CC0 Creative Commons)

- **Concrete**: *Concrete* is the most regularly used human-made material on Earth for construction works. It is one of the major construction materials used extensively while its uses concern various structural applications. It is a composite material, which is formed mostly of Portland cement, water and aggregate, such as gravel, sand or rock. When these materials are assorted together, they create an operable paste which then slowly but surely becomes hard over time. There are several different concrete types such as: ordinary, lightweight, reinforced, and many more.
- **Cast stone**: *Cast stone* is a masonry product made of concrete imitating the appearance of natural-cut stone. It is employed in architectural and construction applications. Cast stone has been the primary building material for several hundreds of years. The known use of cast stone is dated back to France many centuries ago. It can be formed by a particular process from white or grey cements, fabricated or natural sands, crushed stone or natural gravels, and colored with mineral coloring pigments. This happens in order to reach the desired hue and appearance while maintaining resilient physical properties. Cast stone can take the place of any natural building stone.

4.5 Building Pathology

According to Watt (2007), building pathology is a holistic approach to understanding buildings which demands a detailed knowledge of how buildings are designed, constructed, used and changed, and the various mechanisms by which their material and environmental conditions can be touched by an external factor. It is an interdisciplinary approach and needs a broader identification of the ways in which buildings and people respond and react to each other. In addition to that, building pathology and diagnostics focus on the condition or process of degrading and tumble of existing structures. The assessment of the building condition can lead to the identification of the source's defects and the intervention options in a practical and comprehensible form (Harris 2001).

Building pathology provides a good picture on the status of the building. It enables us to quantify the existing damages and designate the appropriate measures. In order to create this visualization for building condition, documentation and analysis are necessary to understand the roots of each damage.

Documentation and modelling help us to understand how the damages react on the overall process of building pathology and provide the overall picture, because it is essential for supporting decision-making. Employing single or fusion of techniques such as photogrammetry, laser scanning, thermal imaging, etc., we seek to create the holistic picture for the status of the building.

4.5.1 Water Damage

Water is the element that drives the growth of mold and bacteria (Fig. 4.6). Exposure to water-damaged buildings has not only direct consequences in its structural condition but causes many health effects. There are many ways buildings become a toxic mixture of microbes, fragments of microbes, etc. Buildings can further progress the growth of fungi, bacteria and mycobacteria as a result of construction shortcomings, incomplete basements exposed to ground water conditions, poor provision of fresh air, inadequate building design, and many more.

Water and moisture can damage a building. There are many ways to do that including flooding, bad weather events, water pipes and sewage overflows. Water can damage historic infrastructure consisting of individual structures and objects, as well as outstanding objects of art placed singly or steadily attached as an integral part of buildings. All these elements are subjected to miscellaneous forces and actions because of water-based activity.

The main moisture source is originating from the environment. Buildings are subject to several sources of moisture from both sides of the building, inside and outside. Depending on a number of factors, such as age, status of the building edifice, an assortment of environmental factors, etc., one or more of the moisture sources can be show up.

Fig. 4.6 Extensive moisture damage in a historic building (*Source* Pixabay/CC0 Creative Commons)

Moisture in masonry is a big issue, which could be summarized as follows (Rosina et al. 2018):

- Monitor the presence of moisture during seasonal cycles.
- Measure the amount of moisture and individuate possible damage thresholds depending on the specific sensitivity of the building and of special decorative elements, such as mural paintings and stucco works.
- Remove the desirable amount of moisture, picking the removal techniques by considering the optimum balance in between effectiveness and invasiveness.

4.5.2 Freeze and Thaw

The are many places across the world where air temperature falls down to minus (−) degrees Celsius during the winter. Under these conditions, the historic buildings' outside walls deteriorate due to the damage from frost. An attentive study of the decay of historic buildings and their environmental conditions is necessary to conclude to the optimum measures. The temperature and humidity in the external and inner side of the historic building as well as the water content of the wall surface should be measured. In general, the main cause of deterioration is usually the damage in the outer parts. The roof is a typical example for that. In practice, it allows rainwater and snow-melt water to penetrate the inner parts of the building (Ishizaki and Takami 2015).

Rainwater is penetrating into the stone mostly through the deteriorated part of the roof and rainwater gutters. In addition to that is the penetration through the deteriorated components of the external walls. The freeze and thaw cycles of the water in these areas are responsible for the deterioration of the building materials during the winter time. Usually, as a protective measure, we prevent water penetration into the building materials in order to keep away frost damage in the winter (Ishizaki and Takami 2015).

4.5.3 Salt Damage

One of the most serious damages to historic buildings is caused by something entirely insidious; salt (Fig. 4.7). It is carried from one place to another by wind and water droplets and is even found in several building materials. Salt is a forceful mineral that may be the reason for a building's façade collapse.

Salt damage affects porous building materials, such as brick, concrete, limestone and sandstone. The consequences of this effect occur when salt crystallized and situated inside of a building material's pores. It derives enough force to cause it to break. Salt damage has strongly impacted many historic structures around the world, such as the marble Grecian monuments, the Egyptian pyramids and many more.

Fig. 4.7 Salt damage to a wall (*Source* Wikimedia Commons/CC3 Creative Commons)

Several building materials already contain salt, such as cement, which is a type of concrete that contains calcium and other alkali sulfates. For instance, through atmospheric pollutants, environmental forces can generate salt damage in buildings. Buildings that are located near the sea encounter similar issues due to saltwater spray throughout the year. There are also other possibilities how salts can enter into constructions: from air pollution, chemical reactions and decompositions, reconstructions and renovations, etc.

4.5.4 Biological Decay (Rotting, Plant Growth...)

Biological decay corresponds to materials' colonization by means of vegetation, organisms and micro-organisms. It also includes the effect by other organisms such as birds and animals nesting, among other biological sources. It also has a mutual relationship with the properties of the substrate and notably with its porousness, in the absence of which the organisms could not be established. In addition to that, the presence of moisture in the surfaces permits organisms to be developed.

The absence of ordinal cleaning and maintenance contributes to the development of damages originating from biological decay. It is possible that damages could pervade several centimeters into the layer of material substrate, as well as into joints and cracks. This type of damage could show up in concrete, iron, masonry and wood.

4.5.5 Corrosion

Corrosion (Fig. 4.8) of historic structures pertains to any process involving the deterioration or degradation, for instance in the case of concrete or metal components. Documentation and condition assessment of such structures is necessary as it can provide extremely useful information with respect to the current condition of the structure and the factors conducing to the corrosion damage. Furthermore, monitoring and assessment can lead to an estimation when the structure may be exposed to further material loss.

In the case of metal components, the best-known situation is that of the rusting of steel. Under normal conditions, in nature, corrosion is an electro-chemical process. As such, it has substantial attributes of a battery. In the corrosion systems very often associated with historic buildings, there may frequently be solely a single metal involved, with water including some salts in solution as the electrolyte. It is possible that corrosion may even occur with water, considering that oxygen is also available.

Metal components used in buildings can be organized in four general classes: those used on the exterior of the buildings, those formed into the construction of the buildings, those used for utilities and services, and finally, those buried under the buildings. For example, metals used on the external elements of the buildings are experienced mainly to atmospheric conditions. The main atmospheric factors have

Fig. 4.8 Corrosion on a building façade (*Source* Pixabay/CC0 Creative Commons)

an effect on the corrosion of metals are temperature, pollution by sulphur dioxide and chlorides. In addition to that, it is the time period during which metal is exposed wet to water.

4.5.6 Cracks

Normally, a crack is a break that becomes visible between neighboring elements. In fact, it is a deviation in form that can be regarded as a deformation. A crack is formed when pressure acting on structures prevails over the material strength, and thus the parts between cracks could display deformations.

The severity of a crack varies and, in practice, it can be described in terms of its (alphabetical):

- Depth
- Direction
- Location
- Pattern
- Width

Moreover, a crack can be diagonal, horizontal, random or vertical. Generally, it can be categorized as either active or non-active. The differentiation of a crack as well as the identification of its behavior in terms of its type is of high importance. An active

Fig. 4.9 Cracks on masonry, stone and brick (*Source* Pixabay/CC0 Creative Commons)

crack is related to a non-stabilized condition that can augment and congest damage or failure of a structure. On the contrary, a non-active crack remains invariable.

A crack could appear on masonry, stone and brick (Fig. 4.9) and concrete. The crack configuration in masonry and concrete is a graduated process, starting with extremely small cracks and ending on large-scale cracks.

4.6 Historic Building Information Systems

In general, an information system is a complete package of components for data collection, storage and processing, and for making available information, knowledge, and digital products. The main elements of information systems are computer hardware and software, databases, telecommunications, human resources and also procedures. The hardware, software, and telecommunications compose the so-called Information Communication Technology (ICT).

A Geographic Information System (GIS) is a computer-based information system for performing geographical/spatial analysis. GIS has four interactive components: the input subsystem for data digitalization, the storage and retrieval subsystem, the subsystem responsible for the geographical/spatial analysis and the output subsystem for digital products (maps, tables, etc.).

Any specific information system, like the historic building information system (HBIS), aims to support specific tasks, management and decision-making.

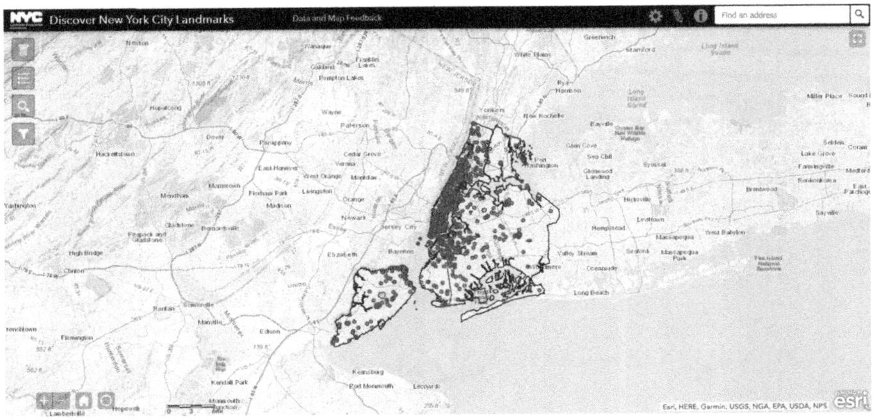

Fig. 4.10 The NYCLPC Web GIS application (*Source* NYCLPC)

A HBIS has the historic buildings at the center of its conceptual and physical implementation.

The New York City Landmarks Preservation Commission (NYCLPC) is the mayoral agency responsible for protecting and preserving NYC architecturally, historically and culturally significant buildings and sites. The Commission was established in 1965, and since then has designated over 36,000 buildings and sites, including 1,405 individual landmarks, 120 interior landmarks, 10 scenic landmarks, and 141 historic districts and extensions in all five boroughs (Manhattan, Brooklyn, Queens, The Bronx and Staten Island) (NYCLPC 2018).

The NYCLPC developed a HBIS as a Web GIS application Fig. 4.10 to make data on all historic buildings more accessible with an interactive map and also make it easier for the public to explore the city's wide range of designated historic buildings. The application includes detailed information on the historic buildings located in the historic districts. The application contains tools to search and filter historic district building data. The users can search and filter by characteristics such as architectural style, architect, building type and era of construction.

4.6.1 The Example of Casa Italiana in New York

The history of the Academy goes back to 1927 when *Casa Italiana* was established at Columbia University as the chair of the Italian Department and as the principal center for Italian Studies in the United States. The building is a neo-Renaissance palazzo built by McKim Mead & White. The Columbia University building is located in 1151-1161 Amsterdam Avenue between West 116th and 118th Streets in the Morningside Heights neighborhood of Manhattan, New York City, and nowadays houses the Italian Academy for Advanced Studies in America (ITALIANACADEMY 2018).

Casa Italiana landmarked in 1978 and represents one of three buildings in the Columbia University campus having achieved the same status. In 1982, the building was also added to the National Register of Historic Places (Dolkart and Postal 2009; White et al. 2010).

The following information is provided by NYCLPC for this specific landmark:

Landmark	Casa Italiana I LP-0991
Landmark type	Individual Landmark
Designation date	3/29/1978
Address	1151-1161 Amsterdam Avenue
Borough	Manhattan
Architect(s)	McKim, Mead & White; William M. Kendall
Style	Renaissance Revival
Construction date(s)	1926–1927
Designation report	http://s-media.nyc.gov/agencies/lpc/lp/0991.pdf.

A snapshot from the NYCLPC Web GIS application, targeting *Casa Italiana* and the area around Columbia University, is provided in Fig. 4.11.

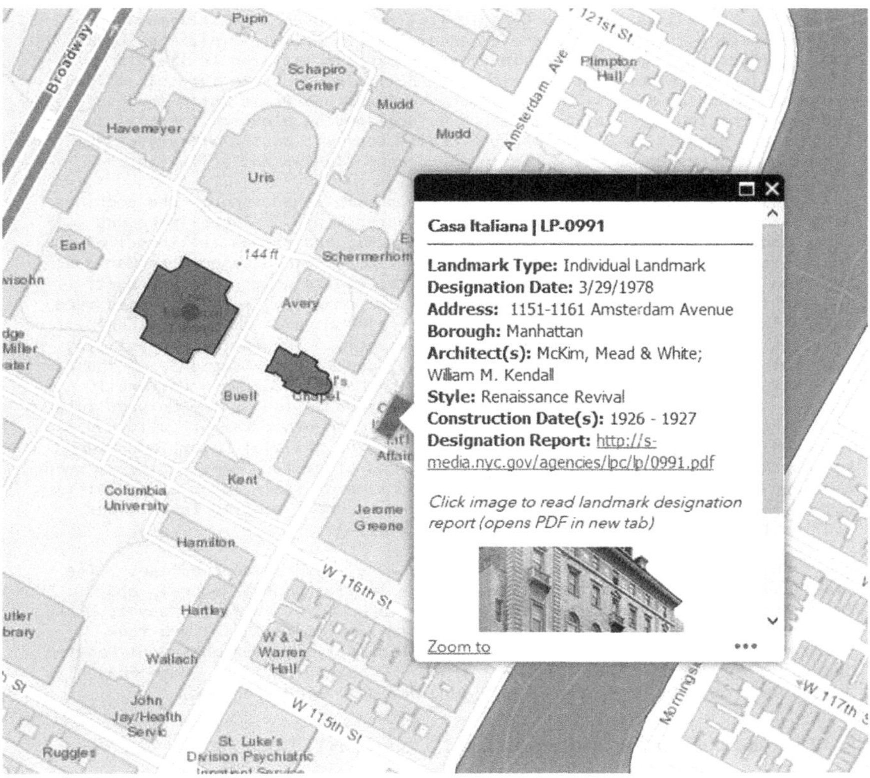

Fig. 4.11 *Casa Italiana* in the NYCLPC Web GIS application (*Source* NYCLPC)

Landmarks Preservation Commission
March 28, 1978, Designation List No. 114
LP-0991

CASA ITALIANA, 1151-1161 Amsterdam Avenue, Borough of Manhattan. Built 1926-27; architects McKim, Mead & White.

Landmark Site: Borough of Manhattan Tax Map Block 1961, Lot 37.

On March 14, 1978, the Landmarks Preservation Commission held a public hearing on the proposed designation as a Landmark of the Casa Italiana and the proposed designation of the related Landmark Site (Item No. 11). The hearing had been duly advertised in accordance with the provisions of law. One witness spoke in favor of designation. There were no speakers in opposition to designation. Columbia University has expressed its support of the designation. The Commission has also received several letters in support of designation.

DESCRIPTION AND ANALYSIS

Casa Italiana, a neo-Italian Renaissance style institutional building, was designed as a center for Italian studies by the prestigious architectural firm of McKim, Mead & White. Built in 1926-27, it is located on the Morningside Heights campus of Columbia University at Amsterdam Avenue and 117th Street.

Casa Italiana had its beginnings in 1914 when a group of students of Italian descent organized an Italian club at Columbia College in an effort to promulgate knowledge of modern Italy. In 1920 a group of 25 students formalized themselves into the Circolo Italiano. Under the chairmanship of student Peter Riccio (later a professor of Italian at Columbia), they began seeking funds to establish a collectionof books specializing in Italy's contribution to culture and learning and facilities to house the collection. President Nicholas Murray Butler encouraged the students in their endeavor and suggested that an Italian House be established as a center for Italian studies and activities within the university context, as the already-established Maison Française was a center for French studies and activities.

The construction of Casa Italiana was accomplished through the combined efforts of Americans, Italians, and Italian-Americans . Fundraising was begun under the leadership of Judge John Freschi, and the establishment of the Institute of Italian Culture at Columbia in 1923 with Professor John Gerig as chairman gave further impetus to the Casa Italiana campaign. The project aroused the enthusiasm of the Italian-American community, and committees were formed throughout the United States to help raise money. Many prominent citizens contributed time and money to the effort. The well-known building contractors, Joseph Paterno, Michael Paterno, and Anthony Campagna, of the firm Paterno Bros., Inc., offered to build the Casa. Columbia University donated the site on the condition that the Casa be suited to the University's needs and conform in general style and structure to other campus buildings. The architectural firm of McKim, Mead & White, the official Columbia University architects, agreed to offer their services at cost. Many of the workmen who were of Italian descent volunteered a portion of their labor as a contribution to the Casa Italiana. The Italian government donated furnishings, and the Italian royal family gave two oil paintings.

The cornerstone was laid on August 5, 1926, by President Butler. The Italian Ambassador to the United States, the Italian Consul General, and representatives of Italian societies in New York City were also present. Construction proceeded rapidly, and the dedication of Casa Italiana took place on Columbus Day, October 12, 1927. The Italian government sent Senator Guglielmo Marconi as their representative. President Butler, Judge Freschi, Joseph Paterno, and Professor Dino Bigongiari delivered speeches.

Fig. 4.12 *Casa Italiana* designation report, p. 1/5 (*Source* NYCLPC)

When opened the Casa was placed under the administration of the Institute of Italian Culture with Professor Gerig as director. In 1930 Giuseppe Prezzolini, who had served as a visiting professor of Italian in 1929-30, was appointed director of Casa Italiana. Durning his ten-year administration, he established a close relationship between the Casa and the Columbia Italian Department, and he furthered the role of the Casa as a center of Italian culture in the United States. Casa Italiana now serves as the home of the Columbia University Italian Department, which has trained more Ph.D.s in Italian studies than any other university in the United States. The Casa also sponsors numerous conferences, seminars, congresses, and forums on a wide range of cultural subjects. Its large auditorium offers a site for many varied musical and theatrical performances. The Casa is a center for interdepartmental studies and events in such fields as art, science, architecture, and medicine, contributing to the broader understanding of Italian influence in these areas.

Casa Italiana is situated outside the boundaries of the original campus master plan which the architectural firm of McKim, Mead & White had designed in 1894-1903. In 1894, Charles F. McKim (1847-1909) had begun to lay out the Morningside Heights campus of Columbia University between Amsterdam Avenue and Broadway and between 116th and 120th Streets. McKim's design was chosen for the university because of its monumentality and its classical style. The preference for this style reflected the success of the Chicago World's Fair of 1893 whose buildings had also been designed in the classical style. The axiality and symmetry of the Columbia University design were in the tradition of the Ecole des Beaux-Arts in Paris, where McKim had studied during the 1860s. In 1903 the firm designed an extension of the campus southward to 114th Street. The extended campus, covering the six blocks between Amsterdam Avenue and Broadway and between 114th and 120th Streets, constituted a new comprehensive design that incorporated and respected the original one of 1894. In the 1920s the university began building outside the extended campus plan. Although not part of the master plan, Casa Italiana maintains the cornice line of the original campus buildings, and the design was praised at the time of construction for conforming to the previously established campus architectural scheme.

The Casa as designed by the firm of McKim, Mead & White is a free adaptation of an Italian Renaissance palazzo. William Mitchel Kendall (1856-1941) was the partner in charge of the design. A graduate of Harvard and the Massachusetts Institute of Technology, Kendall also studied independently in France and Italy. He joined the firm of McKim, Mead & White in 1882 and worked very closely with McKim. Kendall became a full partner in the firm in 1906.

A handsome structure, the Casa Italiana rises six stories on a rusticated limestone base which is pierced by grilled window openings and the main entrance on Amsterdam Avenue. Handsome wrought-iron entrance doors are set within a rich enframement with foliate molding. The smooth limestone facade is enhanced by quoins delineating the end bays. Prominently featured at the second story are full-height arched windows with keyed enframements. These open onto the Casa auditorium which seats approximately 400 people. Handsome stone balustrades accent the window bases; that at the central window projects and rests on the cornice above the main entrance. The third and fourth stories are treated visually as a unit, separated from those stories above and below by band courses. At the third story, arched windows with keyed enframements and balustrades indicate the location of the Charles Paterno Library. Dr. Paterno, the brother of contractors Joseph and Michael Paterno, donated a library collection worth $15,000 , to the Casa when it opened. Three of the rectangular windows at the fifth story are accented by pediments and projecting balconies. Decorative panels flank the rectangular windows at the sixth story which is surmounted by an impressive entablature. A low hipped tile roof rising above this is pierced by two dormers flanking a loggia. The treatment of the facade on 117th Street is similar at each story to that on Amsterdam Avenue.

-2-

Fig. 4.13 *Casa Italiana* designation report, p. 2/5 (*Source* NYCLPC)

The building continues to effectively serve the functions for which it was designed. The largest portions of the interior are occupied by the auditorium and the library. Other floors contain meeting rooms and offices. Casa Italiana remains an adornment to the Columbia University campus in scale and design and an impressive symbol of the institution it houses.

FINDINGS AND DESIGNATION

On the basis of a carefulconsideration of the history, the architecture and other features of this building, the Landmark Preservation Commission finds that Casa Italiana has a special character, special historical and aesthetic interest and value as part of the development, heritage, and cultural characteristics of New York City.

The Commission further finds that, among its important qualities, Casa Italiana is a striking neo-Italian Renaissance style institutional building; that it was designed by the prestigious architectural firm of McKim, Mead & White to harmonize with earlier buildings on the Columbia University campus designed by the firm; that the construction of Casa Italiana was accomplished through the combined efforts of many Americans, Italians, and Italian-Americans; that Casa Italiana serves as the home of the Columbia University Italian Department and as a center for many Italian cultural activities; and that in the scale and elegance of its design, Casa Italiana is a handsome element of the Columbia University campus and an impressive symbol of the institution it houses.

Accordingly, pursuant to the provisions of Chapter 63 of the Charter of the City of New York and Chapter 8-A of the Administrative Code of the City of New York, the Landmarks Preservation Commission designates as a Landmark Casa Italiana, 1151-1161 Amsterdam Avenue, Borough of Manhattan, and designates Tax Map Block 1961, Lot 37, Borough of Manhattan, as its Landmark Site.

-3-

Fig. 4.14 *Casa Italiana* designation report, p. 3/5 (*Source* NYCLPC)

The five-page designation report (Figs. 4.12, 4.13, 4.14, 4.15 and 4.16) prepared by the NYCLPC in March 1978, describes and analyzes *Casa Italiana* and then concludes with the findings, i.e. "*that Casa Italiana has a special character, special historical and aesthetic interest and value as part of the development, heritage, and cultural characteristics of New York City*", and the designation of *Casa Italiana* as Landmark of the City of New York.

BIBLIOGRAPHY

Andrews, Wayne. "McKim, Mead & White: New York's Own Architects." New-York Historical Society Quarterly Bulletin, 35 (1951), 87-96.

Architectural Forum, 127 (July-Aug. 1967) 69-72.

Casa Italiana, Columbia University, Fiftieth Anniversary , 1927-1977. New York: Columbia University, 1977 (pamphlet).

"Casa Italiana, Columbia University." The American Review, II (July-Sept. 1959), 8-14.

["Casa Italiana."] Architecture and Building, 59 (November 1927), 351-352, pl.217.

"Casa Italiana, An Italian Palace in New York." Good Furniture Magazine, 29 (1927), 128-129.

Columbia University. Casa Italiana, 2 boxes. Columbiana Collection, Low Library, New York.

Howson, Roger. Historical Survey of the Casa Italiana. Unpublished typescript. New York: Columbiana Collection, [1940].

McKim, Mead & White. Casa Italiana, Columbia University, New York. Box C-611, 1950, Tube 333. McKim, Mead & White Collection, Print Room, New-York Historical Society, New York.

New York Herald, August 1, 1926.

New York Herald Tribune, January 9, 1926; May 31, 1946.

New York Times, July 29, 1923; Nov. 28, 1924; Jan. 25, 1925; Dec. 10, 1925; Jan. 20, 1926; May 24, 1926; July 18, 1926; March 13, 1927; Oct. 13, 1927; April 24, 1930; Aug. 31, 1930.

New York World, August 6, 1926; August 8, 1928.

Passanti, Francesco. "The Design of Columbia in the 1890s, McKim and His Client." Journal of the Society of Architectural Historians, 36 (May 1977), 69-84.

Prezzolini, Giuseppe. The Case of the Casa Italiana. New York: American Institute of Italian Studies, 1976.

Roth, Leland M. "McKim, Mead & White Reappraised." A Monograph of the Works of McKim, Mead & White, 1879-1915. New edition, four vols. in one. New York: Benjamin Blom, 1973, pp. 11-57.

Withey, Henry F. and Elsie R. Biographical Dictionary of American Architects, Deceased. Los Angeles: Hennessey & Ingalls, Inc., 1970.

Conversation with Francesco Passanti on the McKim Plan for Columbia University, March 1978.

-4-

Fig. 4.15 *Casa Italiana* designation report, p. 4/5 (*Source* NYCLPC)

CASA ITALIANA
1151 AMSTERDAM AVENUE

DATE: 1926-27
ARCHITECT: McKim, Mead & White

Fig. 4.16 *Casa Italiana* designation report, p. 5/5 (*Source* NYCLPC)

References

Cu (2018) Copper Development Association. http://copperalliance.org.uk. Accessed 23 May 2018

Dolkart A, Postal M (2009) Guide to New York City landmarks, 4th edn. Wiley, New York, p 195. ISBN: 978-0-470-28963-1

Flandro X, Thomas-Haney HM (2015) A survey of historic finishes for architectural aluminium 1920–1960. APT Bull - J Preserv Technol 46(1):13–21

Harris S (2001) Building pathology: deterioration, diagnostics, and intervention. Wiley, Hoboken, p 672. ISBN: 978-0-471-33172-8

ICOMOS, Australia (2013) The Burra Charter. Australia. http://australia.icomos.org/publications/burra-charter-practice-notes. Accessed 03 Feb 2018

Ishizaki T, Takami M (2015) Deterioration of the wall of a historic stone building in a cold region and measures to protect it. Energy Proc. 78:1371–1376

ITALIANACADEMY (2018) Columbia University, The Italian academy for advance studies in America. http://italianacademy.columbia.edu. Accessed 06 Feb 2018

Jordan JW (2005) Lime mortar and the conservation of historic structures. Aust J Multi-Discipl Eng 3(1):35–42

NPSa (2018) Preservation briefs. https://www.nps.gov/tps/how-to-preserve/briefs.htm. Accessed 02 March 2018

NPSb (1988) 17 - Architectural character-identifying the visual aspects of historic buildings as an aid to preserving their character. https://www.nps.gov/tps/how-to-preserve/briefs/17-architectural-character.htm. Accessed 02 March 2018

NYCLPC (2018) New York City Landmarks Preservation Commission. http://www1.nyc.gov/site/lpc/index.page. Accessed 06 Feb 2018

Rosina E, Sansonetti A, Ludwig N (2018) Moisture: the problem that any conservator faced in his professional life. J Cult Herit 31(Supplement):S1–S2

Sickels-Taves LB, Allsopp PD (2005) Lime and its place in the 21st century: combining tradition, innovation, and science in building preservation. In: International building lime symposium, Olrando, Florida

Smallman RE, Ngan AHW (2007) Physical metallurgy and advanced materials, 7th edn. Butterworth-Heinemann, Oxford, UK. ISBN: 978 0 7506 6906 1

Wagner DB (1996) In: Brill EJ (ed) Iron and steel in ancient China, 2nd edn. Leiden, The Netherlands. ISBN 90 04 09632 9

Watt D (2007) Building pathology: principles and practice, 2nd edn. Blackwell Publishing, Oxford, p 305. ISBN: 978-1-4051-6103-9

White N, Willensky E, Leadon F (2010) AIA guide to New York City, 5th edn. Oxford University Press, New York, p 496. ISBN: 9780195383867

Winter J et al (2014) Introduction to stone in historic buildings: characterization and performance. In: Stone in historic buildings: characterization and performance, London, UK, pp 1–5. https://doi.org/10.1144/SP391.10

Chapter 5
Planning: Prior Building Surveying and Documentation

Abstract Planning prior to any work related to the surveying and documentation of cultural heritage is necessary to designate the requirements and specifications of a project. A clear identification of requirements and specifications, a timeframe and within an agreed budget is essential. This chapter also discusses issues related to the agreement and the contract, as well as how to choose between 2D and 3D historic building surveying. A well-organized surveying and documentation project will help a team to focus on the project aims, map the implementation, and formulate data collection and analysis activities. In every work, a clear answer on how to preserve all the records is necessary. Safety is also discussed in this chapter, as teams are working both in indoor and outdoor conditions, sometimes using expensive and heavy equipment. Another important issue discussed in this chapter is matters related to standardization.

5.1 Building Surveying Assignment

The assignment of a project like the surveying of a historic building is a very serious job. It is not only the responsibility for delivering a complete project on time and covering all the specified requirements. It is also the heavy, historical and emotional load to manage responsibly the history of the building: what it was and how its existence will continue in the future. Before commissioning such a project, the client, public or private, must clearly designate the requirements and specifications of the project. The contractor should have the skills, knowledge and expertise of carrying out the project entirely, covering all the clearly identified requirements and specifications, on time and within the agreed budget.

In any case, the contractor should have the appropriate and relative experience. In most of the circumstances, the contractor forms a multidisciplinary team consisting

E. Stylianidis, *Photogrammetric Survey for the Recording and Documentation of Historic Buildings*, Springer Tracts in Civil Engineering,
https://doi.org/10.1007/978-3-030-47310-5_5

of professionals from various disciplines, such as architects, civil engineers, conservators, surveying engineers, even historians or software engineers, and many others, depending on the project's extent, type and budget. It is essential that the contractor is competent to understand the overall wishes, requirements and specifications the client set. The client and the contractor should have a good collaboration and interaction in order to establish a smooth project implementation. An attentive briefing, carefully prepared and covering entirely the project's nature, needs, extent, time chart, and budget, is an advantage for the project. It is beneficial for the parties. Nevertheless, this is not always the case.

5.1.1 Agreement and Contract

The contractor will be selected either by direct selection and assignment of the project, otherwise through a tender procedure which follows specific workflows. Definitely, the latter is the typical pipeline used in the public sector or to other authorities related with the public sector. In the private sector, things are more flexible, as there is usually a non-strict public procurement framework.

Once the contractor has been carefully chosen, it is of great significance to reach an agreement and set up the ground on which the project will be implemented. This agreement can be figured out in many different ways, but in any case, it should include legal and formal administration documents such as:

- Appointment agreement
- Conditions of appointment or engagement
- Agreement with subcontractors (if applicable)
- Time chart of the project
- Project method statement and risk assessment
- Bank guarantee (under specific conditions)

By utilizing such type of legal documents which are approved and signed by both parties, a fair agreement will be established. The extent of the services to be provided within the project must be unambiguously defined. All parties should be aware of their duties and also their responsibilities in this mutually acceptable agreement.

During the implementation of the project, there is a possibility an extra and required service or task is not covered by the contract. It is also not covered by an assessment form as it is common to identify such types of inconsistencies or omissions between the conceptual design of a project and its actual implementation. In this case, it is requisite to consort with a brief for this specific service or task, explaining all the details and the relevant requirements and the specifications. This also may require further investigation in this specific field or just a general condition assessment. External consultation or advice could also be considered, in the cases of very special issues that need a more detailed perspective. An amendment to the contract may be necessary, however, additional services can be covered by additional agreements.

The relationship between the client and the contractor is based on the contractual commitments of both sides. The warranties and conditions of the contract define the rights and obligations of both and may be expressly set by the parties or implied by the law of the country in which the project is taking place and the contract was signed.

5.1.2 Requirements

Requirements are a package of specifications set for a project. It is one of the most important technical parts in assigning a surveying and documentation project for a historic building. On the one hand, the client defines the requirements that should be fulfilled, and in practice explicates in detail the requests and the deliverables. On the other hand, requirements mark the specifications that the contractor has to follow in order to achieve these requirements. There should be a harmonic coupling of the two project components, i.e. the client and the contractor.

The first issue is the formulation of expectations. Developing products and providing services that meet the user or client expectations, is critical to success. Especially nowadays where technology-oriented tools and solutions offer many and different possibilities. Analyzing user requirements is the basis for a user-centered approach, generating products and providing services that meet the user's needs at the highest level. User requirements analysis procure accurate outlines of the requested content, and quality demanded by users, now or in the future.

In order to identify user needs, the user's current views and prospects must be supposed to be the case and result in *products requirements*. The desired results that the users wish to reach and succeed, and the tasks they plan to execute with the assignment of the project must be determined. By recognizing the product requirements, it is easier for all to understand the involved and optimum tasks and why they perform certain activities to be able to deliver the specific products.

As all these should be considered within a predefined timeframe and with reasonable use of human and financial resources. Respect to the legal framework is also a crucial part of *procedural requirements*. Specification of procedural requirements includes all the other categorization apart from the products requirements, such as (indicatively):

- Team members' profiles
- The project needs to be finished within a predefined timeframe
- Project delivery by spending a specific amount of money, i.e. the project budget
- Necessity to use specific equipment
- Respect of specific legislation
- Intellectual Property Rights (IPRs)
- Consideration of security and privacy requirements

The usual case is that the owner, i.e. the client, is able to clearly describe the real needs for such a surveying and documentation project for a historic building.

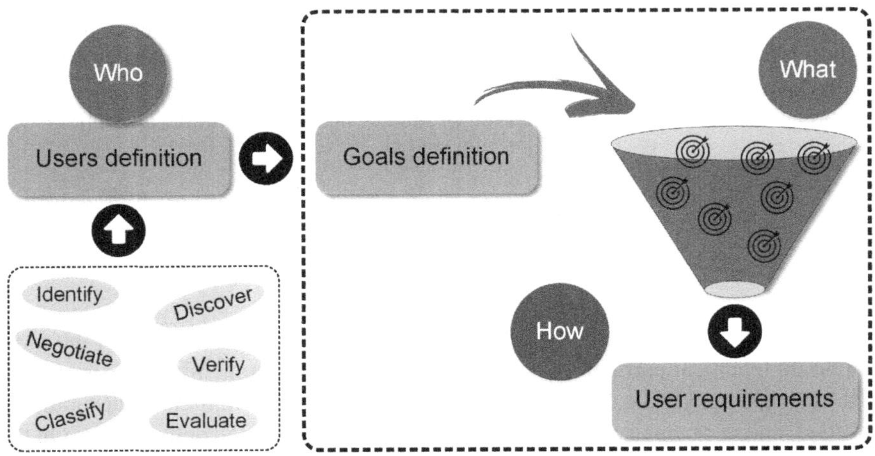

Fig. 5.1 Schematic representation of user requirements workflow

However, there also cases where the owners are not competent enough to define all their real needs, or to define part of the needs, and thus they would very much appreciate the support by experts. Another case is when the owner, even though she/he has the knowledge and skills to describe the requirements, would prefer to hire an independent consultant (individual or firm) to provide professional and scientific support, in order to prepare the user requirements. There are many and different cases, but the user requirements workflow remains more or less the same (Fig. 5.1).

Methods, techniques and tools such as document analysis, focus groups and analysis, interview, observation or questionnaires can be deployed for the extraction of user requirements. In parallel with these, different requirements analysis methods can be set to practical use to complement each other in order to produce more effective conclusions and outcomes. For the execution of the requirement identification and analysis process, a sort of tools can be used in a supplementary way. Typical requirements gathering and analysis methods are listed together in Table 5.1, illustrating their pros and cons, while the particularities of their implementation are featured in the following.

The degree to which user requirements' analysis is successful in the beginning of a project like in the case of surveying and documentation of historic buildings depends to a large extent on two main parameters:

1. The type of the project, i.e. the type of the historic building (size, typology, etc.).
2. The ability of the responsible personnel to respond effectively, regardless the method to be used (Table 5.1).

Collecting the requirements from the users is a very thoughtful piece of work, which requires much effort; the real needs should be fully understandable to all and the risk to fail be small. The personnel that is involved in this process should be

Table 5.1 User requirements analysis tools

Method/Technique	Description	Pros	Cons
Scenarios/Use cases	Elaborate realistic paradigms of how the users may execute their tasks and activities with the historic building surveying outcomes, under a specific preservation context	The uses cases can fetch the user needs to life	The scenarios may increase the expectations of the user too much
User surveys	A collection of written questions directed to a sample of users. The users surveys can help specify the needs, current work practices and attitudes related to the historic building surveying outcomes	It is relatively quick method of defining preferences of large user groups and admits statistical analysis	It does not capture in depth comments and may not allow follow-up
Focus groups	It brings together a cross-section of users in discussion group arrangement. A practical method for the requirements extraction	This technique allows quick abstinence of a broad diversity of user perspectives	Dominant participants in the focus group may sway group disproportionately
Interviews	This is a sequence of fixed questions where the user is able to expand on every specific response	The interviews give the opportunity for fast extraction of ideas and concepts	It is possible to receive different opinions and judgements from different users

well educated and trained, otherwise it will be very hard for them to report the real needs. Especially, the case of historic building surveying and documentation, where the results of such project will most probably lead to the restoration or preservation of the building.

Creativity of the leading persons is required for transposing the user needs into real, clear, well-structured and tangible requirements. For the implementation of a high-level professional work, precise user requirements analysis and specifications are absolutely necessary. If needed, more professionals, experts or technicians should be added in this group to support every aspect of the project.

Task analysis is obligatory for securing the loyal and successful application of the commitments. User requirements analysis is an error prone component of the implementation process and errors not identified at this stage may lead to bigger costs or failures in a later stage. For that reason, user requirements should be demonstrated that are true, accurate and justified, as soon as the overall project design is available.

Creating a survey for the user requirements requires much skill and effort for those collecting the data, for several reasons. When planning a survey, the research objectives must be articulated in such a way as to allow easy and accurate perception or interpretation (Fowler 1995). It needs a detailed vision of the surveying and documentation project, and also knowledge in other disciplines such as psychological sciences. The purpose of surveys or any other tool, which are analyzed below, is very specific. It targets user requirements definition of a surveying and documentation project, the collection of significant amount of focused data about the documentation process itself, the outcomes and their quality.

Scenarios/Use cases: Use case modelling is widely known and used in software development methods as a way of describing the requirements of a software or a system. The adaptation of this methodology to a project like the surveying and documentation of a historic building is what happens here. Thus, the translation of use case modelling from the software development field to a historic documentation and preservation project states that a use case determines a goal-oriented set of interactions between external actors and the outcome to be produced. Actors are parties outside the system that interact with the system. A use case is introduced by a user with a specific goal in mind and finishes successfully when that goal is contented. It describes the sequence of interactions between actors and the project components (e.g. equipment) necessary to deliver the service that meets the expectations and satisfies the goal. A complete set of use cases identifies clearly all the different ways to reach all the goals. Broadly, a use case can be presented through a Unified Modeling Language (UML) use case diagram, in an easy-to-understand structured and narrative. A scenario is an instance of a use case and presents a single path through the use case. Scenarios may be illustrated using sequence diagrams (France and Rumpe 1999).

User surveys: Typical example of this tool towards user requirements gathering is the online surveys. Nowadays, many different options and solutions are available, both commercial, and free and open source software (FOSS). A survey like this, i.e. a web-based questionnaire, is used to collect all the necessary information and the requested user needs. There are tools that even provide direct statistical analysis of the results, provided that the structure of the questionnaire permits. The survey has very fixed and defined goals: prioritisation of user requirements, completeness check of the requirements found through the use of other tools such as the one-to-one and focus group sessions, and confirmation that these requirements are applicable to this project. The analysis of the questionnaire is able to return those results that enable the responsible team to adjust the requirements priorities, e.g. identified as a result from the one-to-one interviews or a focus group session.

Focus groups: Focus groups are rather a typical technique that can provide help in assessing user needs and feelings both before project outcomes and the long-term vision of a documentation project. In a focus group, usually 6–9 users are discussing issues and concerns about the features of the project and the expectations on the outcomes. The group typically endures a few hours and is coordinated by a moderator who maintains the group's focus. Focus groups often unwrap the spontaneous reactions and ideas of the users and picture the group dynamics and organizational matters. In spite of the fact that focus groups can be a powerful tool in such processes, they should not be used as the only source of data.

Interviews: The collection of requirements can also be performed with a number of fairly informal discussions with key personnel that is familiar and experienced with surveying and documentation projects. The personnel should have clear view on the project objectives and the expected outcomes. The one-to-one discussions about the project and the activities of the user to specify the needs, stages and processes for their (daily) activities or their real needs is the core of the interview exercise. The users are asked about the procedures, the systems, the protocols, the products and the data they are using. Inevitably, the one-to-one interviews are more structured with a view to understand whether the research pursued similar procedures and had relevant outcomes or whether it deflected from the initial assumptions that may adopted. These interviews are focusing on more detail about the actual needs due to previous experiences, although they are flexible to investigate new approaches. A typical example in the surveying and documentation projects is the dilemma between analogue (e.g. paper drawings) and digital outcomes, *2D* and *3D* products, etc.

A use case diagram is a dynamic or behavior diagram in UML. In software engineering, UML is a visualization form of a software program, using a set of diagrams. Even though nowadays UML is accepted by the Object Management Group (OMG) (OMG 2018) as the standard for modelling software developing, this knowledge can be transposed in other application fields. Use case diagrams are imitating system functionality using actors and use cases, which are considered as a set of actions, services, and functions that are necessary for system performance. In this context, if a "system" is a product being developed or operated, and "actors" are people or entities operating under decided roles and responsibilities within the system, then this concept can be transposed to any system that involves people who are acting under specific roles to deliver products.

In the case of our interest, there is no software to be developed, but there are products and outcomes to be generated following specific procedures which are bound to clearly defined specifications. Adopting this methodology, a use case/scenario approach can be considered for a project like the surveying and documentation of a historic building or any other project related to the documentation of cultural heritage. As an example, Fig. 5.2 illustrates a use case/scenario for orthoimage production, by using a UML diagram. By following this approach, a package of UML diagrams can be prepared for each procedure and outcome.

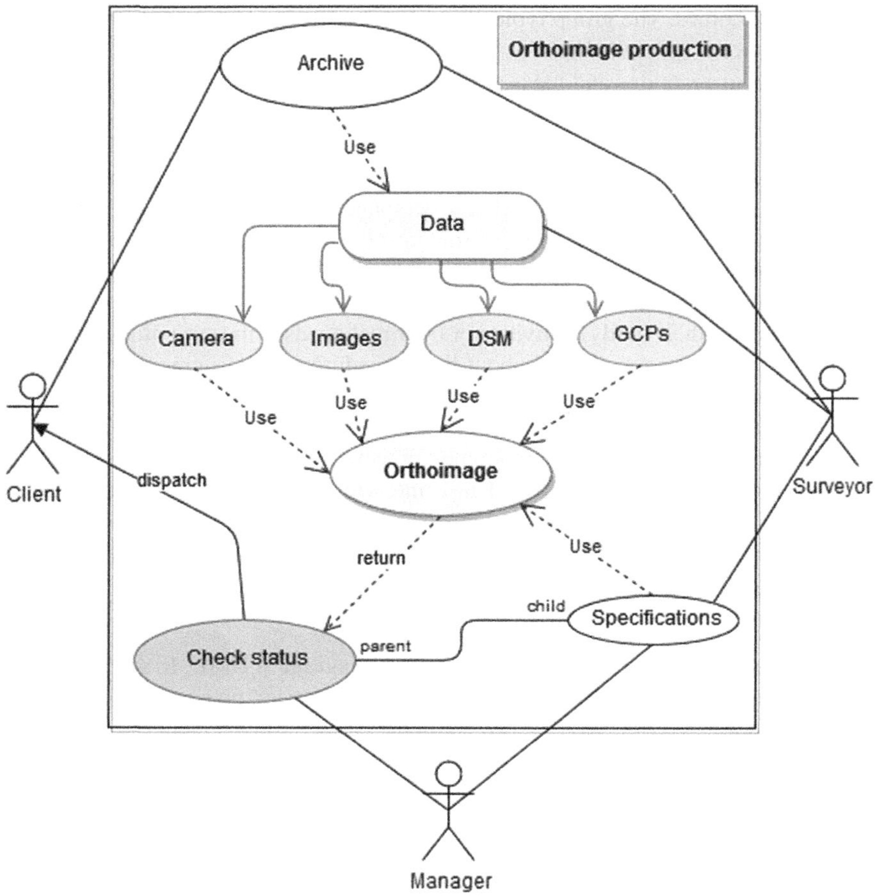

Fig. 5.2 Example of use case diagram for orthoimage production

5.1.3 *Specifications*

Specifications are of high importance for a project and, as the principal means to describe exactly the nature of the procurement requirement, merits particular notice for its important role in project implementation. Especially for complex and interdisciplinary projects, such as surveying and documentation of a historic building which involves people from different disciplines, equipment and new technologies.

Briefly, specifications are the "heart" of the project because they:

- Define the real needs and requirements of the client.
- Numerate the exact products to be delivered by the contractor.
- Set up the quality standards against which bid evaluation, inspection, tests and quality checks are performed.

Specifications can be classified as:

- Functional
- Performance
- Technical

However, it is customary to use the term "technical specifications" to allude to specifications in general. Many times, the three types are merged to explain the requirements with the needful level of details to make certain full perception and organization among parties.

Specifications should most emphasize to all technical and physical details, complemented as necessary by functional and performance specifications, describing exactly the motive and make it clear. It is important to not restrict specifications for products and outcomes to only physical details, especially when supplying accompanying new technologies and systems. Moreover, specifications should be clearly expressed in a generic form, avoiding the use of brand or trade names as much as possible.

In the case of provided services, requests are mostly outlined based on functional and performance criteria, being the first in order of importance specifications to use for services. For instance, requests for services should make available for use background and objectives, the terms of reference (ToR) or statement of works (SoW) required. These include quality standards, the qualifications and experience of personnel required, timeframe, deliverables and outcomes, milestones, reporting, actions for monitoring and evaluation, etc.

Wherever possible, specifications should practice internationally accepted and recognized standards. It is necessary in order to provide a recognized and quantifiable reference for conformity. Such a framework is also necessary because it removes unreliability and gives a clear benchmark the contractors should meet. A characteristic use of standards concerns quality, where quality is related to the perception of the degree to which the services, products and outcomes meet the expectations of the client. Indeed, quality has no specific meaning unless it is related to a particular function. Quality is a perceptual, conditional and rather subjective feature. For that reason, utilizing accepted and determinate standards benefits in elucidating the exact level of quality requested.

In any case, there are different types of projects in surveying and documentation of historic buildings. This reflects to the real needs and the technical specifications. For example, look at the differences between two different but so "similar" projects. A historical building that suffered extensive damage after an earthquake and the same at normal conditions which needs a regular restoration. Even though the building itself is the same, there are different needs and requirements for a surveying and documentation project. There should be different technical specifications on the requested products and outcomes as the form and the particularities of the building features are different.

Heritage England (Andrews et al. 2015) has prepared a publication for metric survey specification in cultural heritage. The publication concerns the supply of survey services based on the options in the specifications that ensure the necessary communication between the information user (the client) and the information supplier (the surveyor). It is the required interaction for a successful metric survey implementation. The publication illustrates a description of the services and standards required for the commission of various types of metric survey.

5.2 2D or 3D Historic Building Surveying?

Documentation is one of the main disposable ways to provide meaning, understanding and reconnaissance of historic buildings. The significance of documentation is considered as an aid to various preservation activities including: recording, monitoring, protection, restoration, conservation, interpretation and management of historic buildings. However, treatment activities on historic buildings, research and application is more and more assisted and determined by digital technology. Documentation tools experienced a significant change over the past years, mainly due to amazing technological developments. As a matter of fact, when dealing with documentation of historic buildings, it is important to represent typology, form, dimensions, materials, colors, decorations, decay and other phenomena related to the building.

The support of various specific skills is often required under an interdisciplinary approach, and it is crucial to select the appropriate tools for an analysis with different professional specializations. Each documentation method has specific advantages and disadvantages. However, due to the continuous rise of digital technology, there is an increasing gap between the specialized technical and non-technical users associated with documentation process. For that reason, the practice of *2D* or *3D* documentation workflows which lead to different approaches, outcomes and products is always a challenging topic among the information providers and the information users.

There are different opinions with respect to what an historic building surveying and documentation project should comprise. This lack of consensus exists mostly between different teams. On the one hand, we have a team that traditionally measures and draws buildings as just a simple surveying project. On the other hand, there is a team that has a real consideration of a heritage structure, such as a historic building. There is a growing difference and a gap that is broadened due to the increasing level of the requested information by the client. It is also the practice of *2D* or *3D* (Fig. 5.3) documentation outcomes, which normally requires different levels skills and analysis of integrated information. Meeting the balance point and the demands set by the client requires additional skills and abilities from the team members carrying out the project. It is not simply collecting site *2D* or *3D* information but collating historical background information as well.

Fig. 5.3 2D (cross-section) and 3D representations of a historic building

5.3 Planning a Surveying and Documentation Campaign

A well-planned surveying and documentation project for a historic building will help a team to focus on the project objectives, map the implementation, and prepare data collection and analysis. Proper planning will result in an effective and efficient surveying and documentation project plan. Gathering and processing the data, the project team will be able to focus on implementing well-supported decisions.

A clearly devised surveying and documentation project plan for a historic building should include the following topics:

1. **Definition**: A detailed proposal for the fundamentals of the involved processes and project definition. Specify measurable objectives to determine the degree to which the project is successful. The definition phase could include actions such as:
   - Calculate how long the surveying and documentation process will take place.
   - Decide on how to organize the personnel in the various tasks to be performed.
   - Use the appropriate equipment.
   - Collect data.
   - Organize data processing, products and outcomes delivery.
   - Succeed in a smooth project implementation based on all contractual obligations.
2. **Value**: It depends on several factors, including:
   - The clear definition of the decisions the project team needs to implement.
   - The relative cost of making an error, based on those decisions.
   - The amount of ambiguity the project will reduce with respect to the historic building preservation.

   This stage of planning will provide assistance to outline the importance of the project and justify the costs.
3. **Cost**: Certainly, no one wants to exceed the agreed budget, nor to find out that the project activities and expenditures went over budget after the project has been accomplished. In practice, with any such project, there will always be associated costs. There are three major costs that the project coordinator can incur, incorporating costs for:
   - Personnel is the most costly expenditure in the case equipment is available. Personnel is necessary not only for the data collection but also for data processing and delivering products and outcomes.
   - Depreciation costs for the available equipment or cost for renting the proper equipment when necessary.
   - Travelling and accommodation costs if necessary for the staff to be used.
4. **Team**: Identify the human resources needed to complete the various tasks and activities. This is determined by the organization model as well as the difficulty of the surveying and documentation.
5. **Timeline**: Making a project timeline will help arrange into a structured order the list of tasks necessary to smoothly complete the surveying and documentation project and delegate those tasks to specific people in the project team. A timeline will help maintain control over the entire project duration for each one of the activities. Developing a plan will drive and coordinate the tasks necessary to start and complete the project successfully. Each project is unique, and the utilization of a proper planning tool will help the project team to better organize the project and succeed the optimum results.

5.4 How to Preserve the Records?

Before starting a surveying and documentation project of a historic building, there should be a clear answer on how to preserve all the records. By default, such a project is considered important, but perhaps even more serious is the preservation of all records. This applies not only to the information and data collected but also to all products and outcomes. Securing all records implies:

- Take into account both the analogue and digital forms of all the records in order to use the appropriate means.
- Easy to perceive naming and identification of both collected information and products/outcomes.
- Selection of an appropriate database management system (DBMS) for archiving reasons.
- A second backup is always available in a separate place, other than the official one.
- Use of appropriate metadata forms and description for every single material of the record.
- Preparation of an accompanied report to record all procedures followed and products/outcomes.
- Secure that every part of the record should be traceable by any user in the future.

5.5 Safety

Safety is important when working on any job or project; however, for some specific jobs, it is a matter of the utmost importance. Teams that are working on outdoor spaces or buildings with many different colleagues, sometimes expensive and heavy equipment are at the top of this list.

Following safety procedures in a surveying and documentation project of a historic building is key to the well-being of all team members and all other colleagues working on the project. Such teams work on many different types of outdoor projects or on historic buildings that may have unstable surfaces. They need to be conscious and aware of specific safety hazards of every job so they can take the appropriate measures and precautions beforehand.

In every project, the team leader should always make mention and point out to the client or site owner's health and safety policy. For any existing risk assessment and any specific hazards, the contractor should have knowledge in order to take the necessary measures.

The project teams often spend time by road, close to fast moving traffic, for instance in the case where the historic building is located in a city center. The drivers can often become inattentive if they quickly and unexpectedly see a person from the corner of their eye. This can and does many times end up in traffic collisions and can be dangerous for people working nearby. There are safety procedures to pursue and people working under these conditions are advised to place safety signs and cones in the proper locations.

It is not unusual for project teams working in places like abandoned historic buildings to come into contact with insects, spiders or even snakes. In order to prevent against unpleasant circumstances, the project team members should wear the proper clothing. In addition to that, a first aid kit as well as a bite kit is a must.

Unstable surfaces constitute another serious threat to the safety and health of the team members. Especially, the case of historic buildings suffering from structural damages, which represent another potential source of danger for personnel. When working in such an environment, personnel should be briefed on the safety procedures. The staff should wear hard hats while their eyes' protection should also be seriously considered.

Any outdoor activity carried out under high temperatures and hot climates represents a number of health dangers, such as dehydration and sunburn to name a few. In order to protect the team members from the sun's effects, they should wear hats, proper clothing, use sunscreen and drink plenty of water. Regular breaks in the shade should be considered as well, while umbrellas can also be used.

5.6 Issues of Standardization

Standards and standardization are based on two main pillars (Fig. 5.4). On the one hand, it is the Quality Assurance (QA), while on the other it is the Quality Control (QC). According to the European Commission and the Joint Research Center (JRC)—ISPRA (EC-JRC-ISPRA 2007):

- "*QA is a set of approaches which is consciously applied and, when taken together, tends to lead to a satisfactory outcome for a particular process. A QA system based on these guidelines will employ documented procedural rules, templates and closely managed processes into which various checks are built. Quality controls (QC) and quality audits are important checks within a QA system.*"
- "*A QC (or check) is a clearly specified task that scrutinises all, or a sample, of the items issuing during, or at the end of, the geometric correction process in order to ensure that the final product is of satisfactory quality. The scrutiny involves review, inspection or quantitative measurement, against well defined pass/fail criteria which are set out in these guidelines.*"

Documentation as a process for planning, acquiring, processing and storing all data and information collected and processed contains rules, guidelines and standards. As data and information collection is the main concern of a surveying and documentation project, unavoidably processes and outcomes should involve standardization issues.

At the international level, there is an absence of one and common standardization framework in the field of cultural heritage documentation. Over the years, there have

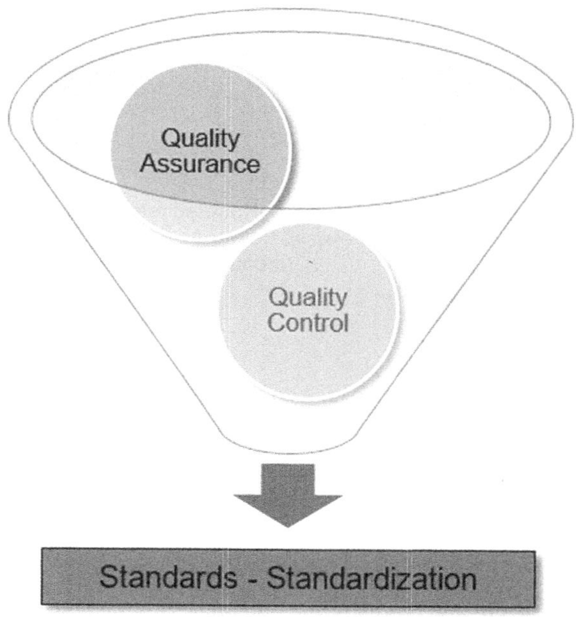

Fig. 5.4 Quality assurance—quality control—standardization

been various initiatives that have laid the concerns and considerations of the scientific and professional community; however, there has never been a single global initiative to resolve this issue.

Data resources are many, data types are also a lot, the heterogeneity of data collection methods varies, and technological developments are changing the working environment. This mixture makes the standardization framework hardly achievable. In addition to that, an international scientific umbrella should cover the whole process for global acceptability.

Standardization is the process of working together to the same end on the development of technical specifications based on consensus. This is achieved on a voluntary basis by incorporating all the key stakeholders from academia, industry, public authorities and other interested parties.

Standards have a profound effect on success for many reasons, amongst and most important, we recognize:

- Provide people and organizations with a basis for mutual understanding.
- They are used as tools to facilitate measurement and communication.
- Facilitate interaction between the various users.
- Enabling people and organizations to comply with relevant laws and regulations.
- Provide interoperability between new and existing products, services and processes.
- Form the basis for the introduction of new technologies and innovations.

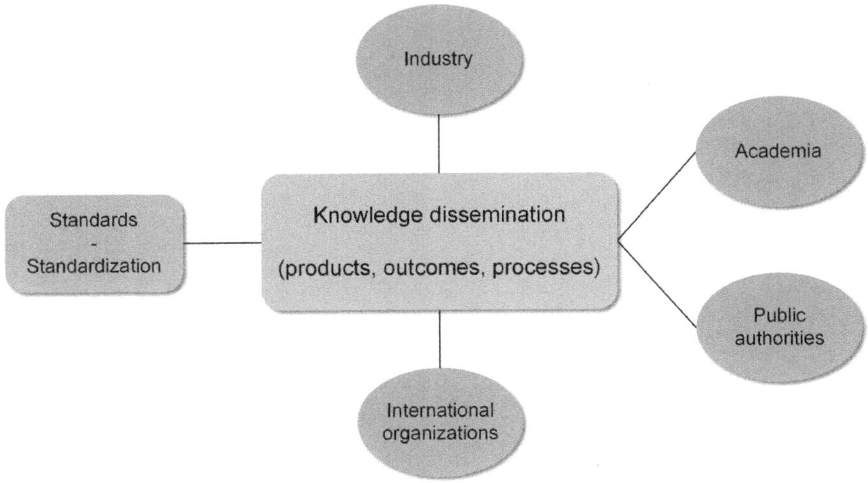

Fig. 5.5 Knowledge dissemination workflows through standards

- Ensure that products and outcomes supplied by different people and organizations will be mutually compatible.

Furthermore, standards also disseminate knowledge in the involved players where products, outcomes and processes supplied by various providers interact with one another (Fig. 5.5).

The use of standards in surveying and documentation projects in cultural heritage domain is very important as it:

- Avoids duplication of work.
- Places the work on a broader foundation.
- Ensures compliance with worldwide acceptable conditions.
- Increases the transparency for prospective users/clients.
- Raises the potential for research funding.
- Can shape a more powerful dissemination strategy.
- Augments visibility.

With respect to standardization for surveying and documentation projects, we put forward for consideration four guiding principles which could be used:

1. Use of common formats
2. Open standards, non-proprietary formats where possible
3. Data interoperability
4. Metadata

However, the introduction of international standards is not an easy work. Various concerns may arise, amongst the most important, we recognize:

- Acceptability by the international community.
- The training of the developing world.
- From international standards to national legislations.
- Technological gaps.
- The Industry 4.0 mindset.

5.6.1 Two International Initiatives

Although none of the initiatives led to international standards, two initiatives originating from CIPA-Heritage Documentation were those activities that have taken place in the past and succeed to introduce new concepts on surveying and documentation procedures.

The first initiative was the "*3 × 3 rules for simple photogrammetric documentation of architecture*" (Waldhäusl and Ogleby 1994), simply known as "*3 × 3 rules*". The motivation of this initiative was the practical use of non-metric cameras for architectural photogrammetry. The authors introduced simple rules, because they are structured in three items with three sub-items each:

- 3 geometric
- 3 photographic
- 3 organizational

Many years later, a newer version of the original one was compiled in a more comprehensive and illustrative way (Fig. 5.6).

The second initiative, RecorDIM initiative, which was presented in a previous chapter, produced two volumes published in 2007:

1. Recording, Documentation, and Information Management for the Conservation of Heritage Places: Guiding Principles (Letellier et al. 2007)
2. Recording, Documentation, and Information Management for the Conservation of Heritage Places: Illustrated Examples (Eppich and Chabbi 2007)

Both publications are available for free at the GCI website.

The first publication, i.e. the guiding principles, reports on 12 principles covering various disciplines, such as design of a project, inventory, selection of method, data types, and institutional accountability. However, these guiding principles do not determine tolerance, performance, or standard for drawing. They are, rather, a framework for the design a documentation project.

In addition to that, RecorDIM formed a dedicated Task Group in 2006 in order to improve, harmonize, elucidate, and combine standards and best practices in the key areas of actual work, technical standards and information management.

PHOTOGRAMMETRIC CAPTURE
THE '3 x 3' RULES*

1 - THE 3 GEOMETRIC RULES

1.1- CONTROL

- Measure some long distances between well-defined points.
- Define a minimum of one vertical distance (either using plumb line or vertical features on the building) and one horizontal.
- Do this on all sides of the building for control.
- Ideally, establish a network of 3D co-ordinated targets or points by a loop traverse around the building.

1.2- WIDE AREA STEREO PHOTOCOVER

- Take a 'ring' of pictures around the subject with an overlap of at least 60%.
- Take shots from a height about half way up the subject, if possible.
- Include the context or setting: ground line, skyline etc.
- At each corner of the subject take a photo covering the two adjacent sides.
- Include the roof, if possible.
- No image should lack overlap.
- Add orthogonal, full façade shots for an overview and rectification.

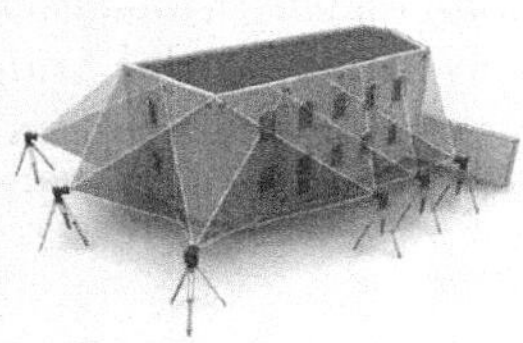

1.3- DETAIL STEREO PHOTOCOVER

Stereo-pairs should be taken:

- Normal case (base-distance-ratio 1:4 to 1: 15), and/or
- Convergent case (base-distance-ratio 1:1 to 1: 15).
- Avoid the divergent case.
- Add close-up 'square on' stereo-pairs for detail and measure control distances for them or place a scale bar in the view. Check photography overlaps.
- If in doubt, add more shots and measured distances for any potentially obscured areas.
- Make sure enough control (at least 4 points) is visible in the stereo image area.

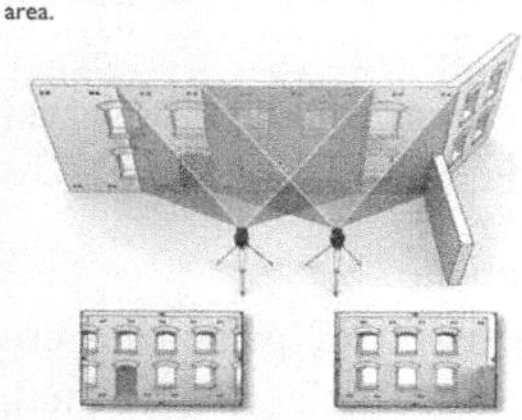

2 - THE 3 CAMERA RULES

2.1 - CAMERA PROPERTIES

- Fixed optics if possible. No zooming! Fully zoom-out. Do not use shift optics. Disable auto-focus.
- Fixed focus distance. Fix at infinity, or a mean distance using adhesive tape, but only use one distance for the 'ring'-photography and one distance for close-ups.
- The image format frame of the camera must be sharply visible on the images and have good contrast.
- The true documents are the original diapositives, negatives or digital 'RAW' equivalents. Set the camera to use the camera its highest quality format.

2.2 - CAMERA CALIBRATION

Use the best quality, highest resolution and largest format camera available:

- 'Medium' format is better than small format. A large sensor is better than a smaller one.
- A wide-angle lens is better than narrow angle for all round photography. Very wide-angle lenses should be avoided.
- Calibrate the camera with a fixed focus lens and tape it there.

- Standard calibration information is needed for each camera/lens combination and each focus setting used. Shooting the calibration screen before capture with each lens will help.
- A standardised colour chart should be used in each sequence of frames.

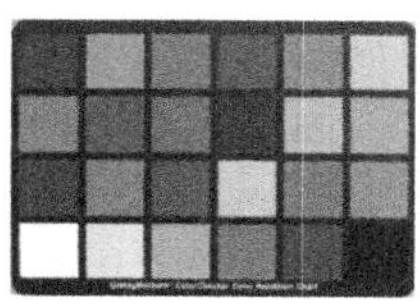

2.3 - IMAGE EXPOSURE

Consistent exposure and coverage is required.

- Work with consistent illumination: beware deep dark shadows!
- Use HDR to capture difficult, unbalanced exposures.
- Plan for the best time of day
- Use a tripod and cable release/remote control to avoid camera movement and get sharp images.
- Use a panoramic tripod head to get parallax-free panoramic imagery
- Use the right media: Black-and-white is sufficient for tracing off lines but colour has some advantages for interpretation and documentation of colours.
- Use RAW or 'high quality' and 'high sensitivity' setting on digital cameras.
- Geotaging the images is recommended.

3 - THE 3 PROCEDURAL RULES

3.1 - RECORD PHOTO LAYOUT

Make witnessing diagrams of:

- The ground plan with the direction of north indicated.
- The elevations of each façade (at an appropriate scale 1:50, 1:100 - 1: 500).
- Photo locations and directions (with frame numbers).
- Single photo coverage and stereo coverage.
- Control point locations, distances and plumb-lines.
- If using 'natural' points a clear diagram showing each point is required.

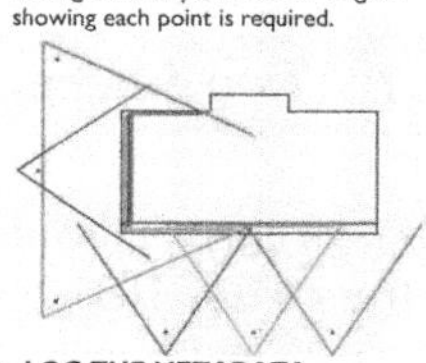

3.2 - LOG THE METADATA

Include the following:

- Site name, location and geo-reference, owner's name and address.
- Date, weather and personnel. Client, commissioning body, artists, architects, permissions, obligations, etc.
- Cameras and optics, focus and distance settings.
- Calibration report, including the geomaetric and readiomaetric results if available.
- Description of place, site, history, bibliography etc.

Remember to document the process as you go.

3.3 - ARCHIVE

Data must be complete, stable, safe and accessible:

- Check completeness and correctness before leaving the site.
- Save images to a reliable location off the camera. Save RAW formats to convert into standard TIFFs. Remember a CD is not forever!
- Write down everything immediately.
- The original negatives are archive documents. Treat and keep them carefully.
- Don't cut into the format if cutting the original film. If using digital cameras, don't crop any of the images and use the full format.
- Ensure the original and copies of the control data, site diagrams and images are kept together, as a set, at separate sites on different media.

*The above text is adapted from a paper presented by Peter Waldhäusl (University of Technology, Vienna, Austria) and Cliff Ogleby (Dept. of Geomatics, University of Melbourne, Australia), at the ISPRS Commission V Symposium "Close Range Techniques and Machine Vision" in Melbourne, Australia, 1994. Simple rules that are to be observed for photography with non-metric cameras have been written, tested and published at the CIPA Symposium in Sofia in 1988.

Fig. 5.6 Photogrammetric capture: the "3 × 3 rules" (*Source* CIPA—heritage documentation)

5.6.2 A Role for Industry 4.0

We are in the middle of a major transformation concerning the manner in which we produce products. This is due to the digitization of manufacturing. This evolution is so captivating that it is being called Industry 4.0 to characterize the 4th revolution that has occurred in manufacturing. In order to reach Industry 4.0, we passed from three other industrial revolutions (Fig. 5.7).

The 1st industrial revolution (Industry 1.0) was marked by a noticeable change: from hand production methods to machines through the use of steam power and water power. The 2nd industrial revolution (Industry 2.0) is also known as the technological revolution. In this case, we had the large-scale use of railway networks, factory electrification, modern production line and the telegraph. There was a clear intention for faster transfer of people and ideas. The 3rd industrial revolution (Industry 3.0), which is also called digital revolution, eventuated in the late 20th century. It was after the end of the two big wars, as a consequence of the industrialization slowdown. In that period, there was an extensive use of computer and communication technologies while machines started to repeal the need for human power.

The 4th industrial revolution will take what was started in the 3rd with the introduction of computers and automation. All these will be enhanced with the use of smart and autonomous systems powered by data and machine learning technology. In practice, what Industry 4.0 does is to optimize the computerization of Industry 3.0.

Inevitably, the fourth industrial revolution will affect surveying and documentation domains as well. Although traditional methods of data collection and imaging will not disappear, it is expected that the key drivers of the Industry 4.0 will be:

- 3D printing
- Augmented reality
- Big data analytics and advanced algorithms
- Cloud computing

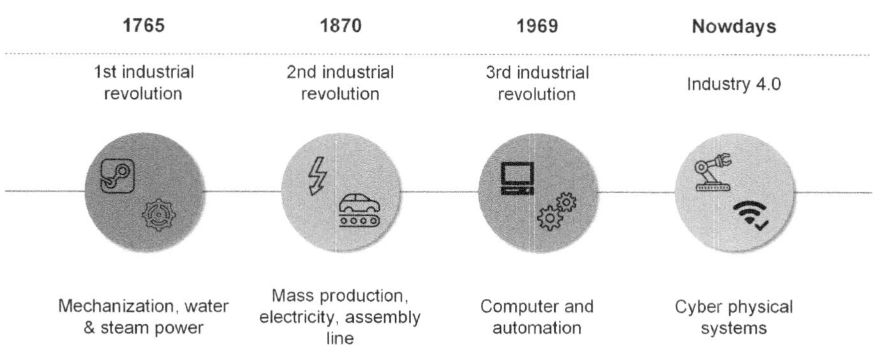

Fig. 5.7 Industrial revolutions pipeline

- Location detection technologies
- Mobile devices
- Smart sensors

will have a decisive influence on data collection, management and representation.

References

Andrews D, Bedford J, Bryan P (2015) Metric survey specifications for cultural heritage, 3rd edn. Historic England, Swindon, 130 pp. ISBN: 978-1-84802-296-6

EC-JRC-ISPRA (2007) Guidelines for best practice and quality checking of ortho imagery. Version issue 2.6. Ispra (VA), Italy

Eppich R, Chabbi A (eds) (2007) Recording, documentation, and information management for the conservation of heritage places. Illustrated examples. The Getty Conservation Institute, 163 pp. ISBN: 978-0-89236-946-1

Fowler FJ (1995) Improving survey questions: design and evaluation. Applied social research methods, vol 38. Sage Publications, London, 192 pp. ISBN: 0-8039-4582-5

France RB, Rumpe B (eds) (1999) «UML»'99: the unified modeling language - beyond the standard, second international conference, Fort Collins, CO, USA, 28–30 October 1999, proceedings. Lecture notes in computer science, vol 1723. Springer, 728 pp. https://doi.org/10.1007/3-540-46852-8. Accessed 20 Feb 2018

Letellier R, Schmid W, LeBlanc F (2007) Recording, documentation, and information management for the conservation of heritage places. Guiding principles. The Getty Conservation Institute, 151 pp. ISBN: 978-0-89236-925-6

OMG (2018) Object management group. USA. https://www.omg.org/. Accessed 16 Feb 2018

Waldhäusl P, Ogleby C (1994) 3 × 3 rules for simple photogrammetric documentation of architecture. In: Fryer J (ed) Close range techniques and machine vision - proceedings of the symposium of ISPRS commission V, vol XXX-5, pp 426–429

Chapter 6
Measurements: Introduction to Photogrammetry

Abstract In terms of content, this is the largest chapter of this book. It is a basic introduction to photogrammetry as a metric technique to perform measurements by means of images. Why and when to choose photogrammetry? Which are the main advantages of using such a technique? A historical review on the invention of photogrammetry is necessary to understand its origin and its evolution across the decades. An introduction to photogrammetry including the basic principles is necessary to understand the 3D measurement technique and its mathematical model, i.e. central projection and the key role of collinearity equations. With respect to 2D, photogrammetric rectification of a single digital image can lead to rectified images, a very useful product in architectural photogrammetry, especially for historic buildings. Interior orientation of the camera is the first step in the photogrammetric process, while exterior orientation is the relationship between image and object space, necessary to calculate 3D coordinates in the object space by measuring points in images. Stereo and multi-image photogrammetry and structure from motion are also discussed in this chapter.

6.1 Why and When to Choose Photogrammetry

The exact location of a point through coordinates, or other metric features, such as areas and volumes, are very important elements in metric sciences or in those cases where a metric procedure should be followed. For example, archaeology is not a metric science, but it needs to calculate or measure points' position in an archaeological site or in an excavation. In the case of historic building preservation, it is necessary to know the exact position in which a beam and a column intersect. Thus, this is the place to elaborate more under which conditions photogrammetry and

E. Stylianidis, *Photogrammetric Survey for the Recording and Documentation of Historic Buildings*, Springer Tracts in Civil Engineering,
https://doi.org/10.1007/978-3-030-47310-5_6

its techniques could be used as an alternative to other techniques, or the conditions when photogrammetry is the only and appropriate solution to measure.

The measurement method is highly dependent on the type of measurements to be performed or on the type of objects to be surveyed. If only a small distance is the requested outcome, there is no need to use any extraordinary technique with high cost equipment or huge processing power. A simple tape and a piece of paper are more than enough. The core of this book, which is the surveying of historic buildings, is a very typical example that is answering the question set. The reasons are many, and amongst them, the potential existence of archival photographs that are presenting the building in another status, the complexity of the structure, the architectural details which cannot be easily captured, just to mention a few.

If the conventional topographic survey is a direct measurement technique, then measurement through images is indirect. In addition to that, performing measurements in images means that all measurements of the object are carried out without any physical contact to the object. In summary, the advantages of using photogrammetry are:

1. **Speed**: In general, data acquisition in photogrammetry, i.e. mainly image acquisition, is not a time-consuming task. This reflects to quick data collection and fast image processing to deliver high fidelity products and outcomes.
2. **Budget**: Considering that nowadays the cost of a digital camera and that quick data collection is feasible, then lower budget can be achieved.
3. **Accessibility**: As already mentioned, photogrammetry is a contactless technique and compared to other techniques (e.g. topographic survey), can be easily applied.
4. **Continuity**: The images constitute a continuous representation of the reality. This means that by using images, an uninterrupted image of the object is captured.

6.2 Historical Review on the Invention of Photogrammetry

It all began when, in September 1858, Albrecht Meydenbauer (1834–1921) (Fig. 6.1), a German architect who was working for the Prussian government, had an accident. Meydenbauer was working as a building surveyor, and one of his first tasks was the documentation of the cathedral in the city of Wetzlar. While he was working for this assignment, he had an unfortunate incident and nearly fell down from the side-aisle of the cathedral. In fact, Meydenbauer himself marked the place where he nearly fell down with an arrow, as illustrated in Fig. 6.2. After this accident, he thought to replace the direct surveying measurements with indirect measurements in photographic images, and thus the principal idea of photogrammetry was born. Meydenbauer is considered the inventor of architectural photogrammetry (Albertz 2001).

After that, Meydenbauer devoted his activities to this idea. In 1860, he prepared a note about buildings' documentation through photography, describing that photographs can store the object (illustrated in photographs) information in detail,

Fig. 6.1 Albrecht Meydenbauer (Albertz 2001)

Fig. 6.2 The cathedral of Wetzlar (Albertz 2001)

Fig. 6.3 Aimé Laussedat (*Source* Wikipedia)

accompanied by high accuracy. He sent this note to the curator of cultural heritage in Prussia, von Quast. Considering the risks for cultural objects, he also developed the idea of a "Cultural Heritage Archive", where important cultural heritage objects should be recorded. He wanted this archive to act as a tool for future reconstruction of the objects in the case of destruction. During the following years, Meydenbauer developed not only a special camera to capture the photographic images but also the photogrammetric methods (Albertz 2001).

As stated by Trinder and Fritz (2008) during the presentation of the historical developments of ISPRS until 2008, in the late 1850s, Aimé Laussedat (1819–1907) (Fig. 6.3) performed the first topographical survey of an area by using a pair of photographs properly distanced from each other. At the same time, Ignazio Porro (1801–1875) (Fig. 6.4) developed the "photogoniometer" and many other ingenious devices. He gave a name to the method, "Metrical Photography", which after further development was later named "Photogrammetry by Intersection". According to Punmia et al. (2005) the concept of using photographs for performing measurements belongs to Laussedat. He was working for the Corps of Engineers of the French Army, and he was the first who used terrestrial images for the production of topographic maps.

Fig. 6.4 Ignazio Porro (*Source* Wikipedia)

In 1851, he produced the first measuring camera, used for these purposes, while he developed the mathematical analysis of photographs as perspective projections.

6.3 Introduction to Photogrammetry

The practical application of photogrammetry in recording and documentation of cultural heritage is more than evident. This chapter discusses and analyzes methods, practices, techniques and tools of photogrammetry used in 3D surveying of cultural heritage. The basic principles of photogrammetry, the equipment, the image acquisition and processing workflows, and the outcomes are amongst the issues discussed in this chapter.

The tremendous growth in ICT, hardware and sensors, and software domain, both commercial and open source, alongside the considerable investments, has made enormous progress and huge increase in the use of photogrammetry in cultural heritage recording and documentation.

Photogrammetry is described by its versatility and applicability in a wide range of application areas. Especially in cultural heritage domain, many and different approaches appeared, from close range photogrammetry to aerial photogrammetry and more recently to UAV photogrammetry. It is not only the type of equipment used in photogrammetry but also the type of objects that it is able to handle, from small museum objects and antiquities, to artefact, archaeological excavations and sites, medium and big size monuments to whole historic towns, to underwater cultural heritage. In this context, this chapter will deal only with the application of photogrammetry in historic buildings.

Even though photogrammetry becomes more and more amicable, attractive, and compatible to the user's needs for user-friendly environments and workflows, the skills of people working with photogrammetry are more than necessary. This is imposed by the ability to design, process, analyze and interpret the products and outcomes originating from photogrammetric procedures and workflows.

In order to practice photogrammetric workflows successfully in historic building recording and documentation, a number of core skills are required in the team implementing such a project (Fig. 6.5). In the next paragraphs, the most important skills are pointed out and analyzed that are necessary in such cases.

Photogrammetry: The team members should have an understanding of the basic principles of photogrammetry, image acquisition and several arrangements in a way that every part of the historic building under documentation is duly covered with an adequate number of images. In order to achieve high fidelity *3D* models, the optimum coverage is required. Photogrammetry with high quality and credibility cannot be performed by considering it as a black box that the user just presses a button and gets an outcome after a short period of time. And if there is no outcome? How about if the result is not the expected one? And if something goes wrong? The team should have the skills, capacity and expertise to handle any poor-quality outcome or hitch.

Image acquisition: The understanding of cameras and other related equipment as the main means for image acquisition is a necessity. This is because it is a precondition in reaching the desired objective of having not only a complete and correct dataset but the appropriate collection of images that can be processed photogrammetrically and obtain the *3D* reconstruction of the building. The knowledge of the camera lens distortions that can have an effect on the quality of the results is also a necessary piece of work to be done for the project team members.

Topography: It is important to understand the proper control and scaling of the project to its real dimensions. Photogrammetry is inextricably linked to topography and the necessary measurements to take place, so as to increase the global accuracy of the products and outcomes. Typically, high-precision topographic measurements at selected points to be used in photogrammetric processes are necessary to ensure not only the control of the generated products but also the required precision of the results. In addition to that, the selected points and their precisely measured coordinates are used for the accurate position of the products in relation to the real world. The use of the topographic and surveying

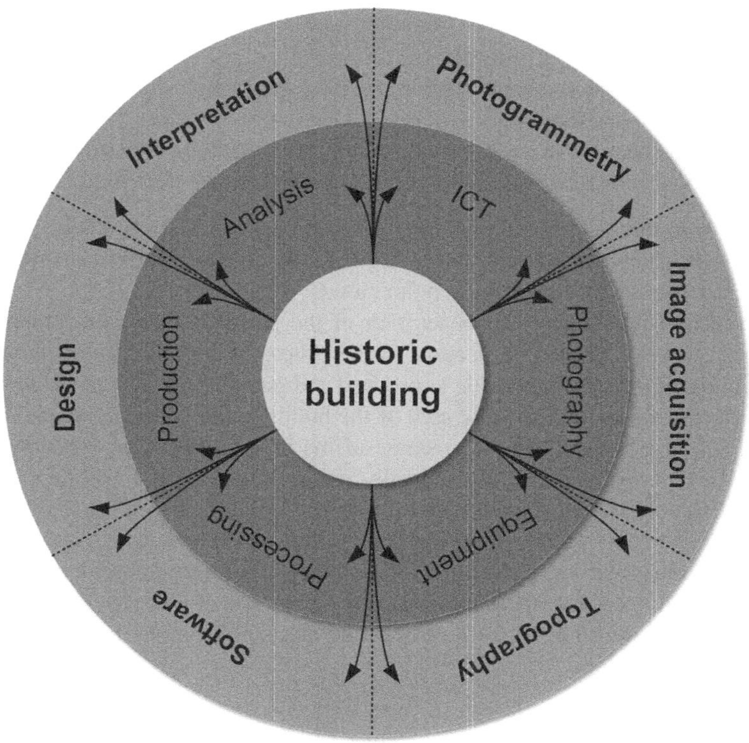

Fig. 6.5 Photogrammetric workflow skills in recording and documentation

instruments are required for the appropriate measurements. The calculating methods and techniques are necessary for the mathematical determination of the points or any other feature coordinates.

Software-Processing: Without data there is no processing, and thus there are no products and outcomes. Data collected during the field work campaign or with any other type of research needs to be processed accordingly to make certain that accuracy and conformity in the application of meeting specific project requirements. It is also important to help make well-informed decisions for the project. Data can be processed only by the suitable software. Apparently, with new software being released on a consistent basis, a number of software solutions and packages exist in the market by different commercial firms. Software concerns all the data type processing chains, from surveying, to photogrammetry, to CAD, image processing, *3D* modelling, etc. The team members should be familiar with the use of the relevant software and consider each project needs. Apart from commercial software, FOSS is freely available for the users as well.

Design-Production: Accomplishing the aim of data processing can lead to justified products and outcomes that are necessary for many and different reasons.

- Firstly, to have measurable, quantitative and qualitative results after a specific production process followed.
- Secondly, for documentation purposes (e.g. building history) that is observed, measured, and mapped.
- Thirdly, because there is a need for concrete products and outcomes that are necessary for further analysis and interpretation (see next step).

Apart from that, additional outcomes could be generated from others' products, for instance, new designs, e.g. a *2D* design of a building façade from an orthoimage or a *2D* cross-section from a *3D* point cloud.

Interpretation-Analysis: The generation of the products does not complete the overall process. There is a need to further interpret and analyze all the results and findings. It is even possible to proceed and generate additional products. For instance, to focus on a specific area of the building and generate supplementary products in a bigger scale. The status of the historic building, especially with regards to its appearance, are typically recorded graphically, although written recording could be combined in such a way as to enhance or emphasize something critical. This is also connected with the condition assessment which is as to a great extent on the effects as on its causes. The condition recording of the causes and the effects of the process of becoming progressively worse will render the required and essential information for a critical and detailed examination of the historic building's pathology.

6.4 Basic Principles of Photogrammetry

Photogrammetry is a *3D* measurement technique where mathematical model utilizes *central projection*, as it is illustrated in Fig. 6.6. But what is central projection and why this is the proper model? It is a projection from one plane to another where the point of the first plane and the image on the second plane lie on a straight line, from a fixed point not on either plane, i.e. the projecting rays are passing through this point which is called the projection center.

So, what is photogrammetry doing? It contains image measurements, processing, analysis and interpretation techniques to derive *3D* coordinates in the object space. This leads to obtaining the location and shape of the building or to the overall *3D* reconstruction of the building. As it is illustrated in Fig. 6.6, two or more images enable space forward intersection, and as a result the measurement of points in the *3D* space. It is obvious that the accuracy of measurements improves if more images are added to the model, since the model becomes more bind and more robust.

However, the transposition from *3D* reality to the *2D* image reflects to a sort of loss of information. Any part of the building that is not displayed in the images is impossible to be measured and thus to be reconstructed in *3D*. This usually happens due to an occlusion problem where parts of the building are in hidden areas. Other cases show that image acquisition was not performed properly to cover every single

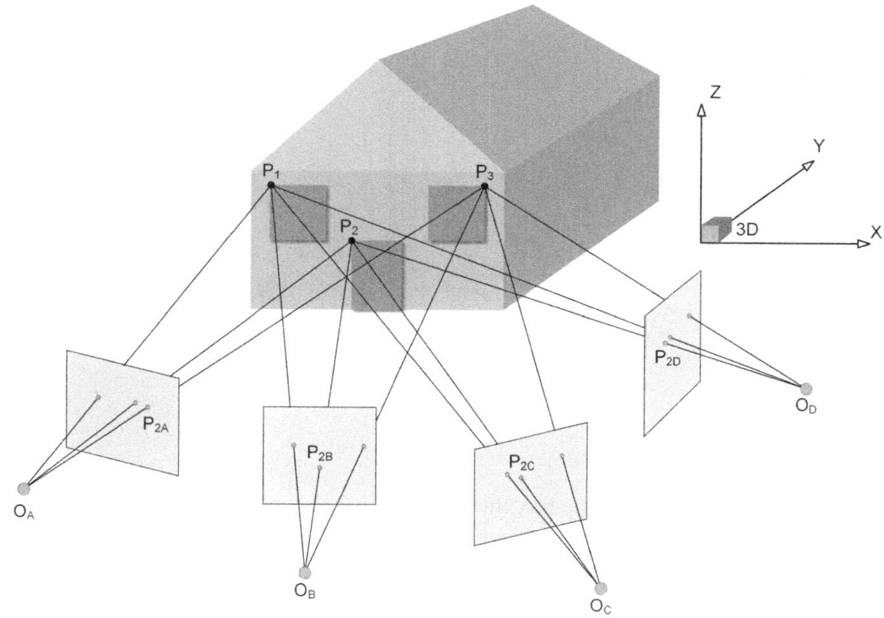

Fig. 6.6 Central projection imaging in photogrammetry

point of the building with more than one image. Another case occurs when parts of the building appear to have very different photometric attributes, and the automatic identification of points cannot be achieved easily.

The position of every single point in *3D* space is defined by its 3D coordinates, i.e. *X, Y, Z*. However, in the image there are only 2D coordinates, i.e. *x* and *y*, as the image is *2D* plane. To better understand the *3D* reconstruction mechanism for an object from images through a photogrammetric process, it is necessary to describe the process of image creation, something which is examined next.

6.4.1 The Pinhole Camera Model

The images captured from a camera do not provide any actual metric size information for every object depicted in the image. Due to the perspective, a small building closer to the camera appears more or less the same as a big building situated at a great distance from the camera. There is no issue if only single imaging is the main part of the job; however, in the case of *3D* reconstruction, the geometry of the represented objects is very important.

The easily understood camera model to capture an image is the pinhole camera, as illustrated in Fig. 6.7. It consists of a box with a hole in one side. The light from

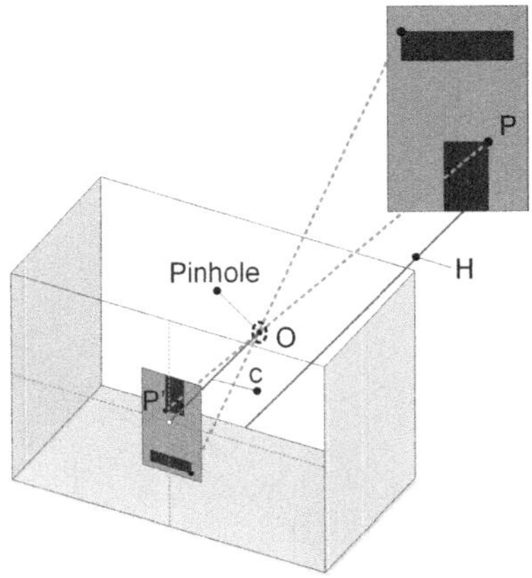

Fig. 6.7 Simple pinhole camera model

the *3D* real world goes into the box through the pinhole and is projected at the back of the box. In this case, the real object is reproduced as an upside-down image, i.e. rotated by 180° as it is shown on that figure.

In addition to that, the image can be projected onto a digital sensor that is part of a digital camera, or a photographic film in a typical film-based camera, or even onto a translucent screen for real time viewing. The projection or perspective center O, through which all image rays are passing, is the most important reference point in the pinhole camera model. Nowadays, digital cameras, more commonly used, can be modelled by using the pinhole camera model. The interior orientation of the camera determines the position of the projection center considered in relation to the image coordinate system. The distance between the image plane and the projection center is called principle distance c, and it is the most significant parameter of camera interior orientation. A critical attribute in this model is the scale of the image. The image scale specifies the way in which two parameters are related, i.e. the object distance H and the principal distance c. In practice, it is a scale factor, and it is given by the following Eq. (6.1):

$$s = \frac{c}{H} \tag{6.1}$$

6.4.2 The Mathematical Model

In order to explain a real phenomenon, collect and analyze data, there is a need to establish one or more mathematical models to describe it. In many cases, these models are based on geometric relations. As already explicated, photogrammetry is using central projection as a mathematical model. Since photogrammetry measures points in images in order to calculate the *3D* coordinates of points in the real environment, there should be a model to describe the connection between the image and the real space. Thus, from the photogrammetric measurement point of view, the following should be known:

- The internal geometry of the camera used for image acquisition, as the measurements are taking place in the images.
- The relationship between the object coordinate system and the camera coordinate system.

Practically, the camera's internal geometry is known if the camera calibration is available (Sect. 6.4.4). This is called, *interior orientation*. With respect to the two coordinate systems, i.e. the object and the camera coordinate systems, the position and orientation of the images is necessary. This is called *exterior orientation* and can be calculated if known points are available in the object space. These points are called *control points*.

The internal geometry of a camera is described by three main parameters, and it is illustrated in Fig. 6.8.

- The camera constant.
- The principle point.
- The corrections to the image distortions.

The camera constant c is almost equal to the camera focal length. Focal length is forming an essential foundation of the photographic lens, and usually it is expressed in millimeters (mm). It should not be confused that it is the actual length of the lens. In fact, it is the measurement of the optical distance from the point where the light rays converge to create the image to the digital sensor or to the 35 mm film -if it is an analogue camera- at the focal plane in the camera. The focal length of the camera lens is determined when the lens is focused at infinity. But what does a number of *35* mm or *150* mm mean for a focal length? The smaller the focal length, the wider the angle of view and thus the lower the magnification. On the contrary, a much higher value of the focal length reflects to a narrow angle of view and a high magnification.

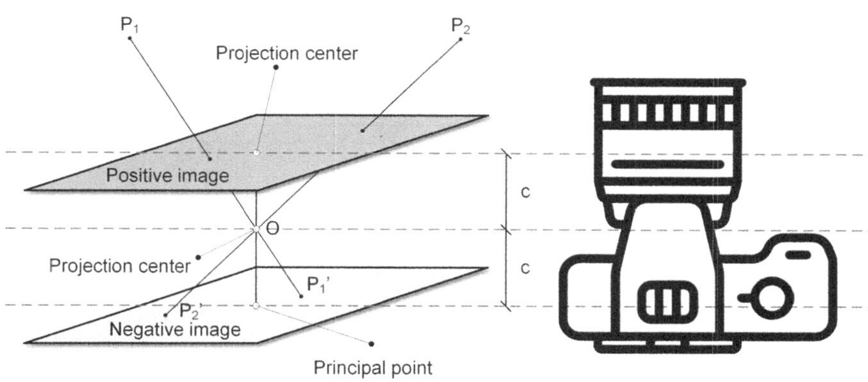

Fig. 6.8 Internal geometry of a camera

In summary, focal length implies how much of the scene will be captured (angle of view), and how large the objects will be (magnification).

The principle point (x_o, y_o) is defined by the intersection of the image plane and a line that is perpendicular to the image plane and goes through the projection center. The principal point as one of the fundamental parameters in camera calibration is extensively discussed by Clarke et al. (1998).

As the mathematical model utilized by photogrammetry for *3D* measurements is central projection, Fig. 6.9 illustrates the connection between a camera, i.e. an image, and the object coordinate system. This means that the object point *P*, the image point *p* and the projection center *O*, lay on the same line. This is the so-called *collinearity condition*. For that reason, straight lines in the object coordinate system show up as straight lines in the image plane, and this is the explanation for the collinearity. The formal coordinate transformation between the object coordinate system and the camera coordinate system is given by the following matrix-vector notation Eqs. (6.2), which are called *collinearity equations*.

$$\begin{bmatrix} x - x_o \\ y - y_o \\ -c \end{bmatrix} = \lambda \cdot R \cdot \begin{bmatrix} X - X_o \\ Y - Y_o \\ Z - Z_o \end{bmatrix} \tag{6.2}$$

The terms appearing in the collinearity equations (6.2) are explained using the following:

x, y	image point coordinates
X, Y, Z	object point coordinates
x_o, y_o	principle point coordinates
c	camera constant
λ	scale factor
R	3D rotation matrix
X_o, Y_o, Z_o	camera projection center coordinates in the object coordinate system.

The collinearity equations (6.2) can be written as group of equations, as expressed by Eqs. (6.3). In practice, with these equations, a known point in the (*3D*) object space (*X, Y, Z*), can be projected into the image plane (*x, y*).

$$\begin{aligned} x &= x_o - c \cdot \frac{R_{11} \cdot (X - X_o) + R_{21} \cdot (Y - Y_o) + R_{31} \cdot (Z - Z_o)}{R_{13} \cdot (X - X_o) + R_{23} \cdot (Y - Y_o) + R_{33} \cdot (Z - Z_o)} \\ y &= y_o - c \cdot \frac{R_{12} \cdot (X - X_o) + R_{22} \cdot (Y - Y_o) + R_{32} \cdot (Z - Z_o)}{R_{13} \cdot (X - X_o) + R_{23} \cdot (Y - Y_o) + R_{33} \cdot (Z - Z_o)} \end{aligned} \tag{6.3}$$

The parameters R_{ij} that appear in the Eqs. (6.3) are the elements of the rotation matrix *R*. In practice, this rotation matrix describes the *3D* orientation of the image with respect to the *3D* object coordinate system (*X, Y, Z*). Each one of the rotation matrix elements is expressed in accordance with the three angles, i.e. ω, φ and κ, which are the rotation angles (of the image) about the X, Y and Z axis respectively (Fig. 6.10). In a matrix expression, R is signified by the following Eq. (6.4):

$$R = \begin{bmatrix} R_{11} & R_{12} & R_{13} \\ R_{21} & R_{22} & R_{23} \\ R_{31} & R_{32} & R_{33} \end{bmatrix} \tag{6.4}$$

Each one of the R_{ij} matrix elements is calculated by the following Eqs. (6.5):

$$\begin{aligned}
R_{11} &= \cos\varphi \cdot \cos\kappa \\
R_{12} &= -\cos\varphi \cdot \sin\kappa \\
R_{13} &= \sin\varphi \\
R_{21} &= \cos\omega \cdot \sin\kappa + \sin\omega \cdot \sin\varphi \cdot \cos\kappa \\
R_{22} &= \cos\omega \cdot \cos\kappa - \sin\omega \cdot \sin\varphi \cdot \sin\kappa \\
R_{23} &= -\sin\omega \cdot \cos\varphi \\
R_{31} &= \sin\omega \cdot \sin\kappa - \cos\omega \cdot \sin\varphi \cdot \cos\kappa \\
R_{32} &= \sin\omega \cdot \cos\kappa + \cos\omega \cdot \sin\varphi \cdot \sin\kappa \\
R_{33} &= \cos\omega \cdot \cos\varphi
\end{aligned} \tag{6.5}$$

The collinearity equations (6.3) express the connection between the image coordinates and the object coordinates. Solving these equations with respect to the object coordinates, the following Eqs. (6.6) are formulated.

$$\begin{aligned}
X &= X_o + (Z - Z_o) \cdot \frac{R_{11} \cdot (x - x_o) + R_{12} \cdot (y - y_o) - R_{13} \cdot c}{R_{31} \cdot (x - x_o) + R_{32} \cdot (y - y_o) - R_{33} \cdot c} \\
Y &= Y_o + (Z - Z_o) \cdot \frac{R_{21} \cdot (x - x_o) + R_{22} \cdot (y - y_o) - R_{23} \cdot c}{R_{31} \cdot (x - x_o) + R_{32} \cdot (y - y_o) - R_{33} \cdot c}
\end{aligned} \tag{6.6}$$

As illustrated in Fig. 6.9 and Eqs. (6.3), one object point reflects to one image point. In addition to that, in Eqs. (6.6) it can be seen that Z coordinates cannot be calculated. This is because in the collinearity of one single image, in each image point there is an infinite number of possible points in the object space, and thus from a single image it is impossible to perform a *3D* reconstruction of the object. In order to calculate the Z coordinates, one or more images captured from a different position and presenting the same object point are necessary (Fig. 6.6).

Before doing so, the transformations devised methodically in Eqs. (6.3) and (6.6) suppose the knowledge of the following independent values:

- (x_o, y_o) principle point coordinates
- c camera constant

As already explained in the camera internal geometry, these three parameters are known as the *interior orientation* parameters. In practice, they fix the center of projection of the *3D* bundle of rays with respect to the image plane.

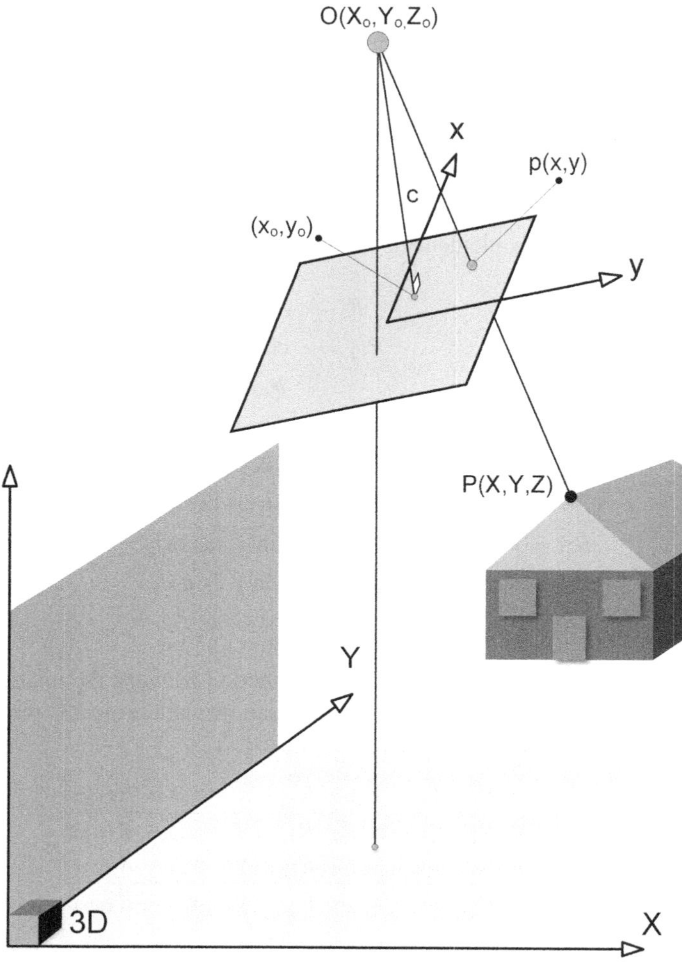

Fig. 6.9 The relationship between the image and the object coordinate system

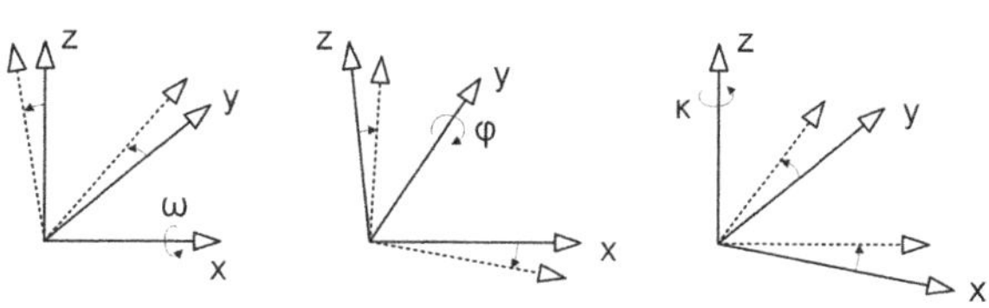

Fig. 6.10 Organization of the three rotation angles ω, φ and κ about the 3 coordinate axes X, Y and Z

- (X_o, Y_o, Z_o) camera projection center coordinates in the object coordinate system
- $(\omega, \varphi, \kappa)$ parameters defining the image rotation with respect to the object coordinate system.

These six parameters, i.e. 3 for the position and 3 for the orientation of the image, are called *exterior orientation* parameters. In practice, they describe exactly the position and settled way of the *3D* bundle of rays in relation to the object coordinate system.

In order to clearly define the central projection of a single image a total of nine parameters is required, i.e. the parameters of the interior and exterior orientation. The values of these nine parameters can be determined in several ways. On the one hand, the interior orientation values are specific for each camera and can be either determined and given by the manufacturer of the camera or can be determined by a process, which is called camera calibration (Sect. 6.4.4). On the other hand, the six parameters of the exterior orientation are normally determined by using control points (Sect. 6.4.5). These are points where both the image and object coordinates are known. According to Eqs. (6.3), if the interior orientation parameters are known, i.e. (x_o, y_o, c), three control points are enough to calculate the parameters of the exterior orientation. For each control point, two Eqs. (6.3) are formulated, i.e. six equations in total, from which the six exterior orientation parameters can be determined. In the case of more control points:

- for 4 control points, there are two couples of (x, y) image point coordinates and thus 8 Eqs. (6.3), and from these 8 equations the 6 parameters of the exterior orientation, i.e. $(X_o, Y_o, Z_o, \omega, \varphi, \kappa)$ can be calculated.
- for 5 control points, there are two couples of (x, y) image point coordinates and thus 10 Eqs. (6.3), and from these 10 equations the 6 parameters of the exterior orientation, i.e. $(X_o, Y_o, Z_o, \omega, \varphi, \kappa)$ can be calculated.
- for 6 control points, there are two couples of (x, y) image point coordinates and thus 12 Eqs. (6.3), and from these 12 equations the 6 parameters of the exterior orientation, i.e. $(X_o, Y_o, Z_o, \omega, \varphi, \kappa)$ can be calculated.
- ... and so on.

In this case, where the equations are much more than the unknown parameters, a least squares adjustment approach is followed to determine the unknown parameters.

6.4.3 Single Image Rectification

Photogrammetric rectification of a single digital image that leads to rectified images are well known in architectural photogrammetry, especially for historic buildings. It is well suited for use to a great extent in building façades. Hemmleb and Wiedemann (1997) give an overview on the different photogrammetric single image techniques, like digital image rectification, but they also extend their study in unwrapping of parametric surfaces and differential rectification methods.

Under specific conditions, it is possible to extract *2D* metric information from a single image. This is the case where the object can be considered as a plane, and thus there is a relation between two planes: the image plane and the plane of the object. If all the object points lie in a plane, this means that there is no elevation and $Z = 0$. In this case, Eqs. (6.6) can be written as:

$$X = \frac{a_1 \cdot x + a_2 \cdot y + a_3}{c_1 \cdot x + c_2 \cdot y + 1}$$
$$Y = \frac{b_1 \cdot x + b_2 \cdot y + b_3}{c_1 \cdot x + c_2 \cdot y + 1} \tag{6.7}$$

It is obvious from Eqs. (6.7) that a single image is adequate for the *2D* reconstruction of a plane object. The eight independent parameters of these equations describe exactly the central projection of a plane object. The eight parameters of this projective transformation can be determined by having known the coordinates of four control points (*X, Y*) and by measuring their image coordinates (*x, y*). There is no *Z* coordinate, as the object is considered flat. If so, then eight Eqs. (6.7) can be formulated, and the eight unknown coefficients $a_1, a_2, a_3, b_1, b_2, b_3, c_1, c_2$ can be calculated.

According to Eqs. (6.7), the original image, i.e. the small image plane, should be transformed to a new image, i.e. the rectified image and the object plane, as illustrated in Fig. 6.11. Apart from the rotation occurring due to the rotation between the image

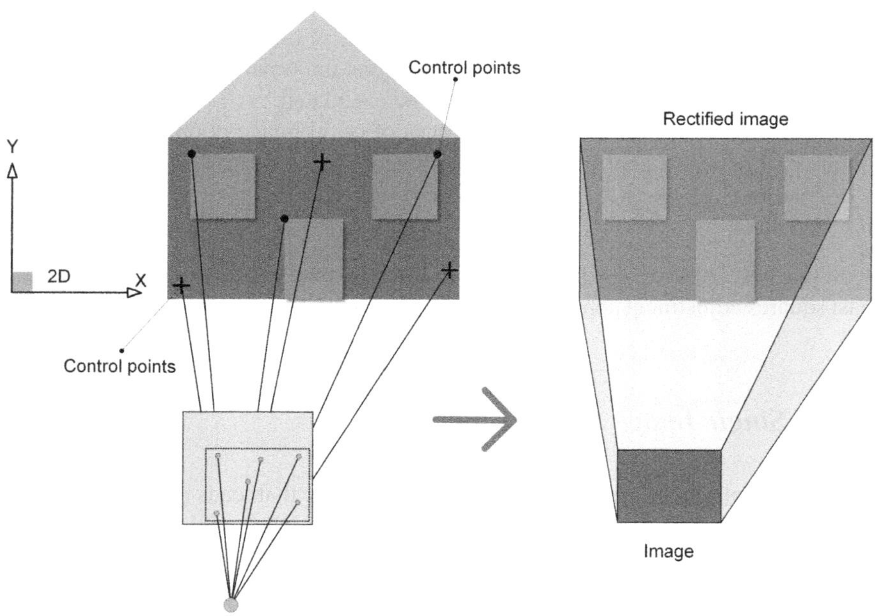

Fig. 6.11 Rectification of a building façade

and the object plane, there is a scaling factor. The production of the rectified image implies that the (small) image is forced to projective geometry and thus the projective transformation and rectification. The outcome of this process is a rectified image, which is a geometrically correct and an undistorted digital image. The transformation of the distorted (original) image to the rectified image is performed with the help of Eqs. (6.7).

As the image matrix is known in the original input image, the (x, y) coordinates of all the pixels in the image coordinate system are known as well. On the other hand, the rectified image is much bigger than the original image, and the pixel values are unknown, as illustrated in Fig. 6.11. The pixel size is defined by the user, e.g. *1* cm, and has a reference in the object coordinate system. As there is not one by one matching between the input and the output image, in order to find the optimum pixel value for each one of the rectified image's pixels, resampling techniques are followed. The three most commonly used resampling techniques are presented next.

Nearest Neighbor Interpolation

The simplest way to assign a pixel value by using a digital image processing resampling technique is the *nearest neighbor method.* In this technique, the pixel value in the original input image is given to that position in the new image matrix of the rectified image that lies nearest, as illustrated in Fig. 6.12.

In practice, this technique concerns an originally empty image matrix for the rectified image in the object coordinate system. The number of pixels in this image matrix will be rather larger than that in the input image. These pixels are transformed into the original image using the inverse transformation:

$$(x, y) = f^{-1}(X, Y) \tag{6.8}$$

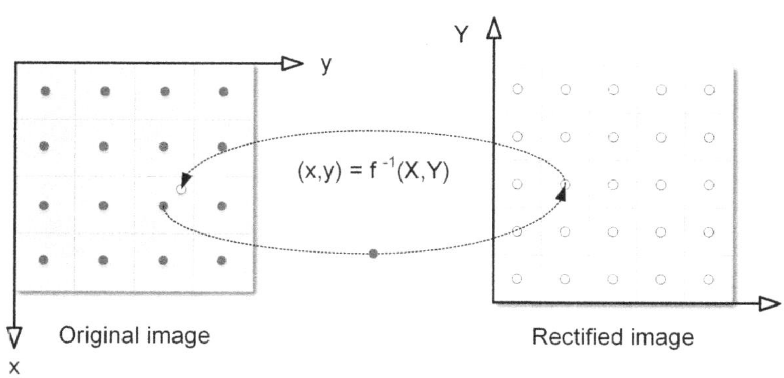

Fig. 6.12 Nearest neighbor approach for image resampling

In this case of the inverse transformation, the specific equations are defined next:

$$x = \frac{(b_2 - c_2 \cdot b_3) \cdot X + (a_3 \cdot c_2 - a_2) \cdot Y + (a_2 \cdot b_3 - a_3 \cdot b_2)}{(b_1 \cdot c_2 - b_2 \cdot c_1) \cdot X + (a_2 \cdot c_1 - a_1 \cdot c_2) \cdot Y + (a_1 \cdot b_2 - a_2 \cdot b_1)}$$
$$y = \frac{(b_3 \cdot c_1 - b_1) \cdot X + (a_1 - a_3 \cdot c_1) \cdot Y + (a_3 \cdot b_1 - a_1 \cdot b_3)}{(b_1 \cdot c_2 - b_2 \cdot c_1) \cdot X + (a_2 \cdot c_1 - a_1 \cdot c_2) \cdot Y + (a_1 \cdot b_2 - a_2 \cdot b_1)} \quad (6.9)$$

The pixel value at that position in the original input image can be found using the nearest neighbor method. Then, this pixel value is transposed into the new and rectified pixel. By following this example for all image pixels, the rectified image is filled properly. In the example provided in Fig. 6.12, and in position $X = 2$, $Y = 3$ of the rectified image, the pixel value calculated at the inverse point displayed within the pixel $x = 3$, $y = 3$ will be recorded.

Nearest neighbor is not the ideal interpolation method while resampling. A disadvantage of the nearest neighbor method is that, in general, it leads to the sawtooth effect, i.e. the resulting image is not smooth enough, and the visual effect may annoy the users.

Bilinear Interpolation

An interesting and rather usable resampling solution is bilinear interpolation. In this case, the requested pixel value is calculated by the values of the four neighboring pixels, as presented in Fig. 6.13.

In bilinear interpolation method, the position of pixel *P(u, v)* is as shown on Fig. 6.13, while its pixel value to be transposed to the rectified image is influenced, and thus is calculated, by the four pixel (neighbor) points *A*, *B*, *C* and *D*. The closest distance to the point *P*, the value is greater, which shows the greater effects. By supposing the coordinates of points *A*, *B*, *C* and *D* as:

- $A \rightarrow (x, y)$

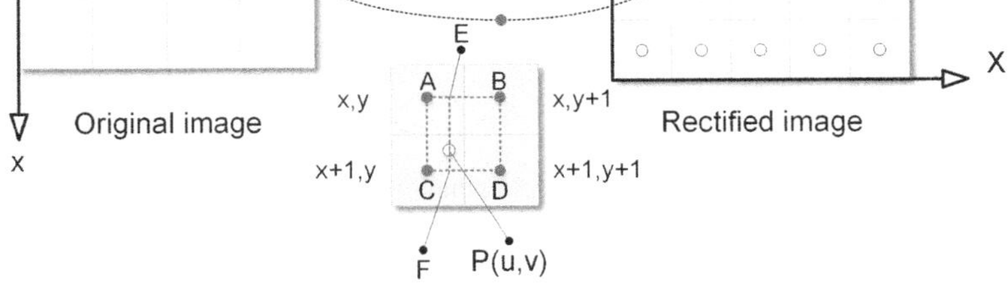

Fig. 6.13 Bilinear interpolation for image resampling

- $B \rightarrow (x, y+1)$
- $C \rightarrow (x+1, y)$
- $D \rightarrow (x+1, y+1)$

the interpolation algorithm in this case consists of three consecutive steps. The model that is followed is considered as next:

$$f(x, y) = \sum_{i=0}^{1} \sum_{j=0}^{1} a_{ij} \cdot x^i \cdot y^i \tag{6.10}$$

$$f(x, y) = a_{11} \cdot x \cdot y + a_{10} \cdot x + a_{01} \cdot y + a_{00} \tag{6.11}$$

Step 1 Calculate the influence of A and B and thus declare point E.

$$f(x, y+v) = [f(x, y+1) - f(x, y)] \cdot v + f(x, y) \tag{6.12}$$

Step 2 Calculate the influence of C and D and thus declare point F.

$$f(x+1, y+v) = [f(x+1, y+1) - f(x+1, y)] \cdot v + f(x+1, y) \tag{6.13}$$

Step 3 Calculate the influence of E and F and thus declare point P which is the ultimate goal.

$$\begin{aligned} f(x+u, y+v) = \{&(1-u) \cdot (1-v) \cdot f(x, y) - (1-v) \cdot v \\ &\cdot f(x, y+1) + u \cdot (1-v) \cdot f(x+1, y) \\ &+ u \cdot v \cdot f(x+1, j+1)\} \end{aligned} \tag{6.14}$$

Bicubic Interpolation

Bicubic interpolation is resampling without being identical to bilinear interpolation. In this case, for the unknown pixel *P(u, v)* in the image, the influence sphere is being enlarged to covering an area of 16 adjacent pixels. The pixel value of P is calculated by these 16 pixels according to their distance to P (Fig. 6.14). Compared with bilinear interpolation, bicubic interpolation extends the influence with more points and utilizes advanced interpolation algorithm.

In order to cover all the 16 values of the adjacent pixels, an interpolated surface is created which is smoother than ones obtained by nearest neighbor or bilinear interpolation. Bicubic interpolation can be performed successfully using either Lagrange polynomials, cubic splines, or cubic convolution algorithm. It provides smoother results, but it is time consuming.

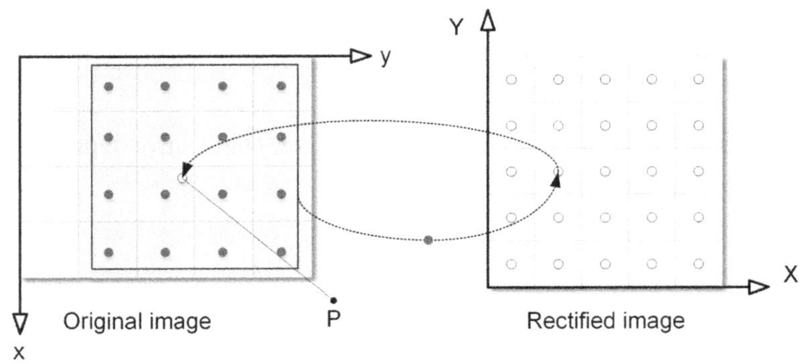

Fig. 6.14 Bicubic interpolation for image resampling

The bicubic model used is like the following:

$$f(x, y) = \sum_{i=0}^{3} \sum_{j=0}^{3} a_{ij} \cdot x^i \cdot y^i \tag{6.15}$$

Single Image Rectification Example
Figure 6.15 illustrates an oblique digital image presenting a building façade. This building façade is considered as plane, and it was captured by using a digital camera Canon EOS 400D. Using conventional surveying techniques, 8 control points measured in the building façade, and their coordinates calculated in a local coordinate system, based on the building façade. The control points are well distributed, covering all parts of the object plane.

A FOSS for digital image rectification, namely VeCAD Photogrammetry, was used to:

- Measure the image coordinates for the 8 control points.
- Calculate the transformation parameters.
- Perform the digital rectification of the input image.
- Generate the rectified image.

VeCAD Photogrammetry software was developed by Tsioukas (2007), and it is based on the source code of VeCAD. This software incorporates full CAD functionalities with a photogrammetric module for digital image rectification of a central projection image. This is possible by using either control point measurements or vanishing point geometry correction.

As the output image, i.e. the rectified image refers to the object coordinate system, a pixel size shall be chosen in meter (m) units. In this case, the selected pixel size for the rectified image was set at the level of *0.01* m, while the overall error estimated for the 8 control points is *0.007* m.

Fig. 6.15 Oblique digital image of a building façade. The positions of the control points are layered over the image

The rectified image is the result of the image rectification process, i.e. the transformation process for correcting the displacements due to the tilt of the image shooting axis and thus to project the image onto the "real" plane in the object coordinate system. As it can be seen in Fig. 6.16, the rectified image is free from the perspective, compared to the original image in Fig. 6.15. All the (structural) lines in the building façade (doors, walls, windows) are horizontal and perpendicular with respect to the object coordinate system.

The data log from the image rectification process is given in Table 6.1. The output of this process provides the control points numbers, the coordinates in the object and image coordinate system, and the errors determined in each control point by the rectification process.

The accuracy of the image rectification depends on the:

- Accuracy of control points measurements in the image coordinate system; the points should be identified and clicked carefully on the exact locations.
- Accuracy of control points measurements; the topographic measurements should be performed following all the typical procedures while the accuracy of the points should be in the area of few millimeters (mm).
- Number of control points; not only 4 points, which is the minimum number, should be captured.
- Position of control points; the points should be well distributed, covering all the building façade.

Fig. 6.16 The rectified image of the building façade

Table 6.1 The data log of the rectification process

ID	X	Y	x	y	Sx	Sy
0	−2.457	3.232	1595.56	523.70	0.0001	−0.0098
1	2.221	3.233	2617.81	586.02	0.0116	0.0040
2	−1.302	1.201	1854.04	949.45	−0.0116	−0.0157
3	1.038	1.211	2397.25	978.33	0.0159	0.0253
4	3.990	0.047	3107.56	1249.26	−0.0179	−0.0132
5	−2.356	−1.100	1566.58	1498.62	−0.0122	0.0163
6	−3.650	−3.860	1116.73	2312.15	0.0079	−0.0048
7	2.513	−3.965	2906.99	2315.36	0.0062	−0.0022

6.4.4 Camera Interior Orientation—Camera Calibration

Interior orientation of the camera is the first step in the photogrammetric process. Without known interior orientation, no further step can be performed. The most important elements of the interior orientation are given in Fig. 6.8. The main reason for which interior orientation is a prerequisite is to define the position of the perspective center and lens distortions. If the interior orientation of a camera is known, then the camera can be called metric camera. An amateur camera is considered as a non-metric camera because the interior orientation changes every time the camera is focused.

As the images are not created by themselves but through the cameras, it is necessary to focus more on the camera. Like in every equipment or instrument, it is needful to know the way it functions. For instance, by using a tape to measure a distance, one should be sure and convinced that the *1* m in the tape reflects to *1* m according to the international standards in the International System of Units (SI). SI is abbreviated from the French Système International (d'unités). The process to check the way an instrument works is called *calibration*.

As in photogrammetry, the real need is results of high accuracy and fidelity, completing a camera calibration is very important. An accurate camera calibration is prerequisite for the extraction of precise and good in quality *3D* metric information from images. Camera calibration is the process of finding the accurate camera parameters that are capturing the images. Therefore, during the camera calibration process, the interior orientation of the camera is determined. The interior orientation parameters describe the metric camera features required for the photogrammetric process. Practically, the interior orientation elements are the following:

1. The principle point coordinates, (x_o, y_o).
2. The principal distance c.
3. The corrections for image distortions Δx and Δy.

Thus, a camera is considered calibrated if the following parameters are well known: principal distance, principal point offset and lens distortion. Usually, principal distance and focal length are terms that cause confusion to the users. The principal distance is equal to the focal length scaled for any magnification or decrease in the print mode where the image measurements are captured or scaled for any change in the image plane location from the infinite focus plane. Two constraints must be fulfilled so as the focal length and the principal distance are identical. First, the camera should be focused at infinity, and second, the image must not be scaled down or magnified.

There are several close-range photogrammetry camera models, but broadly sensor orientation and calibration is carried out with a perspective geometrical model by means of bundle adjustment (Brown 1971). Bundle adjustment makes available a simultaneous determination of all system parameters. Precision and reliability estimations of the extracted calibration parameters estimation are also provided. In addition to that, mutual relationships between the interior and exterior parameters, and the object point coordinates, along with their determinability, can be quantified (Remondino and Fraser 2006).

As a light ray arrives at an angle with respect to the optical axis, it will get in the camera lens and refracted by the camera lens. Then, it will come out at a different angle than the initial one. As the in and out angles are not identical, the system is dissimilar to the ideal central perspective, and thus the points of the object space are not projected in the appropriate position in the image space, due to the refraction in the camera lens system. For the highest accuracies in close-range photogrammetry, these variations of the lens distortions in the image plane should be considered. In this case, Eqs. (6.3) are extended in a way to include the corrections of the image distortions Δx and Δy, and thus the new equations can be written in a new form as following:

$$
\begin{aligned}
x &= x_o - c \cdot \frac{R_{11} \cdot (X - X_o) + R_{21} \cdot (Y - Y_o) + R_{31} \cdot (Z - Z_o)}{R_{13} \cdot (X - X_o) + R_{23} \cdot (Y - Y_o) + R_{33} \cdot (Z - Z_o)} + \Delta x \\
y &= y_o - c \cdot \frac{R_{12} \cdot (X - X_o) + R_{22} \cdot (Y - Y_o) + R_{32} \cdot (Z - Z_o)}{R_{13} \cdot (X - X_o) + R_{23} \cdot (Y - Y_o) + R_{33} \cdot (Z - Z_o)} + \Delta y
\end{aligned}
\tag{6.16}
$$

In photogrammetry, radial and decentering distortions are considered as the two major lens distortions. Additional (distortion) parameters can be added as well.

Radial lens distortion: It is clear that this type of distortion causes radial errors while measuring in the image plane. Even though there are both symmetric and asymmetric radial distortions, usually only symmetric are of high importance due to their magnitude, compared to asymmetric. Thus, the term radial lens distortions concerns the symmetric distortions, which are the result of radially symmetric imperfections and cause variations in lateral magnification with radial distance. Symmetric radial distortions can be either in positive or negative mark while the magnitude depends on the radial distance from the image center. The positive sign in the radial distortions leads the points to be located closer to the image center. On the contrary, the negative sign in the radial distortions is the reason that points are imaged far away from the image center. The first (+) is called pincushion distortion while the second (−) barrel distortion. Both are illustrated in Fig. 6.17.

Decentering distortion: As the camera lens is, in practice, a system of lenses, this distortion happens due to the incorrect arrangement, i.e. the misalignment, of the axes of the camera lenses along a common axis. In addition to that, it is also the misalignment of the normal from the image plane with the optical axis of the camera. The visual representation of decentering distortion is given in Fig. 6.18.

Additional parameters: In this category of additional parameters, failures other than radial and decentering distortions are included. For instance, in-plane distortion refers to deformation effects in the image plane. Even though the integrity of digital sensors is very high, and thus real in-plane distortions are very small, it is possible to occur during image capturing. In addition to that, it is also feasible

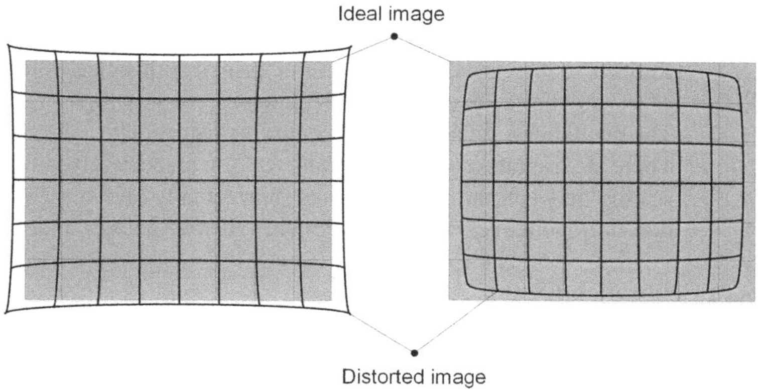

Fig. 6.17 Positive (left) and negative (right) symmetric radial distortions

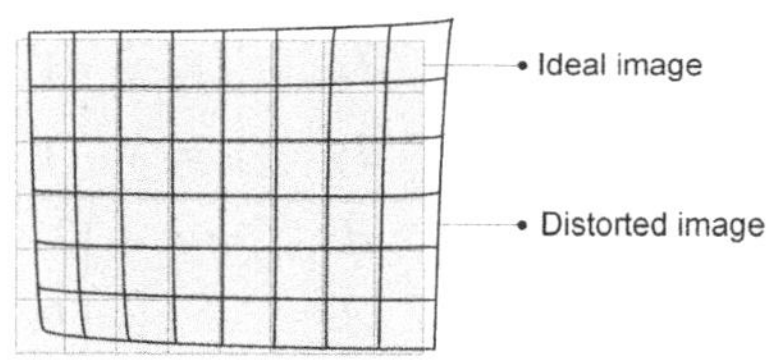

Fig. 6.18 Visual representation of decentering distortions

to add an extra term for the non-orthogonality between the axes. Out-of-plane distortion refers to image plane unflatness.

Various models have been introduced to define the optimum way Δx and Δy could be expressed. An extensive review of camera calibration methods and models has been prepared by Clarke and Fryer (1998). One of the most widely used models is the Brown model (Brown 1971). Fraser (2013), based on the Brown model, is discussing the camera calibration using self-calibration with the aid of coded targets, and he introduced the distortion terms Δx and Δy, like the following:

$$\begin{aligned} \Delta x &= -\frac{\bar{x}}{c} \cdot \Delta c + \bar{x} \cdot r^2 \cdot K_1 + \bar{x} \cdot r^4 \cdot K_2 + \bar{x} \cdot r^6 \cdot K_3 + (2 \cdot \bar{x}^2 + r^2) \cdot P_1 \\ &\quad + 2 \cdot P_2 \cdot \bar{xy} + b_1 \cdot \bar{x} + b_2 \cdot \bar{y} \\ \Delta y &= -\frac{\bar{y}}{c} \cdot \Delta c + \bar{y} \cdot r^2 \cdot K_1 + \bar{y} \cdot r^4 \cdot K_2 + \bar{y} \cdot r^6 \cdot K_3 \\ &\quad + 2 \cdot P_1 \cdot \bar{xy} + (2 \cdot \bar{y}^2 + r^2) \cdot P_2 \end{aligned} \tag{6.17}$$

where r is the radial distance to the image point:

$$r^2 = \bar{x}^2 + \bar{y}^2 = (x - xo)^2 + (y - yo)^2 \tag{6.18}$$

Within the self-calibration model (6.17):

Δc	Reflects to the correction of the initial principal distance value.
K_1, K_2, K_3	These are the radial distortion coefficients.
P_1, P_2	The coefficients reflect to the decentering distortion.
b_1, b_2	There are in-plane correction parameters for, respectively, differential scaling between the pixel spacings (horizontally and vertically) and non-orthogonality (axial skew) between the two axes, x and y.

A Camera Calibration Example

A modified version of Brown model for camera calibration was introduced by Balleti et al. (2014). They developed a free software that offers even non-expert user the possibility to obtain metric data. The camera calibration model introduced in this case uses even order polynomial coefficients to model the radial and the tangential distortion of the lenses (OpenCV 2016), as illustrated in Eqs. (6.19):

$$\begin{aligned} dx_{radial} &= x \cdot (1 + K_1 \cdot r^2 + K_2 \cdot r^4 + K_3 \cdot r^6) \\ dy_{radial} &= y \cdot (1 + K_1 \cdot r^2 + K_2 \cdot r^4 + K_3 \cdot r^6) \\ dx_{tan} &= x + [2 \cdot P_1 \cdot x \cdot y + P_2 \cdot (r^2 + 2 \cdot x^2)] \\ dy_{tan} &= x + [P_1 \cdot (r^2 + 2 \cdot y^2) + 2 \cdot P_2 \cdot x \cdot y] \end{aligned} \tag{6.19}$$

In order to perform a practical example scenario with this software, six images (Fig. 6.19) were captured in a way that the calibration page covers the total width and height of the sensor from different locations. For this calibration example scenario, a *Canon EOS REBEL T2i* with a *55* mm lens was used. The camera uses a Complementary Metal–Oxide–Semiconductor (CMOS) sensor type, the highest still image resolution reaches 5184 × 3456 pixels, while the sensor size is 22.3 mm × 14.9 mm. Consequently, the pixel size of this camera sensor is 4.3 μm.

Usually, the camera sensor dimensions are provided in pixels; however, since the camera calibration parameters (c, x_o, y_o) are expected to be used in measurements, they are provided in millimeters (mm). The remaining camera calibration parameters, according to the chosen calibration model, i.e. K_1, K_2, K_3, P_1, P_2 are expressed in different units. The results of this calibration process provided the results illustrated in Table 6.2.

The calibration estimated accuracy is provided by the OpenCV routine that the software is using. It calculates simultaneously the extrinsic and intrinsic parameters of all the images used during the calibration process. This accuracy is expressed by the average reprojection error, and it is given in pixels. In fact, it is the difference of target points' position defined in the original images from their estimated position using the determined parameters from interior and exterior orientation.

The calibration software delivers a calibration project file with all the appropriate parameters for further use. In addition to that, it also provides a module to use this file in the future to undistort images taken with the same camera modes, as used during this specific calibration.

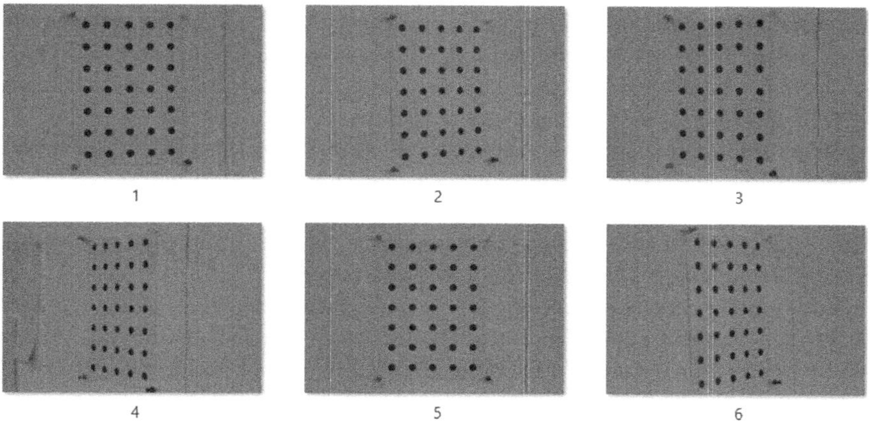

Fig. 6.19 The 6 images captured by Canon EOS REBEL T2i camera for camera calibration

Table 6.2 Interior orientation parameters estimated in the calibration process for the camera Canon EOS REBEL T2i

Parameters	Estimated values
$c(mm)$	51.667
$x_o(mm)$	0.603936
$y_o(mm)$	−0.253952
$K_1(mm^{-2})$	−0.711150
$K_2(mm^{-4})$	103.842613
$K_3(mm^{-6})$	−3624.891602
$P_1(mm^{-1})$	0.003761
$P_2(mm^{-1})$	−0.003413
Average reprojection error (pixels)	3.24

6.4.5 Control Points—Exterior Orientation

The aim of photogrammetry is to calculate *3D* coordinates in the object space by measuring points in images. Consequently, the image and the object space should be connected. Exterior orientation concerns the connection between two spaces: the image and the object. From the practical point of view, this is accomplished by defining the position of the camera in the object coordinate system. The position of the camera is determined by the location of its perspective center as well as by its attitude, something which is expressed by three angles around the three axes.

The Role of Control Points

Despite the advances in sensor technology, control points remain necessary in photogrammetric processes for a large sort of applications that need such features in their procedures. In close-range photogrammetry, image orientation processes require control points. This source of information is necessary for scaling the outputs in ground truth as well as for ensuring the accuracy of the photogrammetric processes. Image orientation is based either on automatic or interactive measurements as the location of the control points in the images is a prerequisite.

When used correctly, control points improve the global accuracy of the photogrammetric outcomes. Namely, control points supply support in order to ensure that (*X*, *Y*, *Z*) coordinates of any point on the outcomes correspond accurately with actual object coordinates. This is of high importance in cases where precise recording and documentation and true global accuracy are required.

Each cultural heritage recording and documentation project is unique, and not all projects demand a high level of global accuracy. For that reason, it is important to estimate the nature of each project separately before deciding to proceed in control points' establishment, measurement and use. However, generally speaking, recording and documentation projects in archaeology and architecture benefit from the use of control points in photogrammetric workflows. The diagram in Fig. 6.20 is a procedural roadmap that could serve as a guide while using control points.

In any measurement, accuracy and precision are of great significance. In order to achieve results of very high fidelity by following photogrammetric procedures, control points should be measured accurately and precisely. However, accuracy and precision are two terms that usually are confused. Although the words are used interchangeably in relaxed and unconcerned conversation, they have not the same meaning. The difference between accuracy and precision is depicted in Fig. 6.21.

Accuracy or geometric accuracy expresses the closeness of a measurement to its true value. It is a "qualitative concept" and that, by nature, a true value cannot be determined. In theory, a true value is that value that would be acquired by a measurement characterized as perfect. Considering that there is no perfect measurement, it is impossible to be aware of the true value. Human inability to carry out perfect measurements and, as a result of that, determine true values does not mean that there is no meaning for accuracy.

Even though accuracy is expressed in qualitative terms such a "good," or "bad", we have the ability to make quantitative measurements as well. This means that we

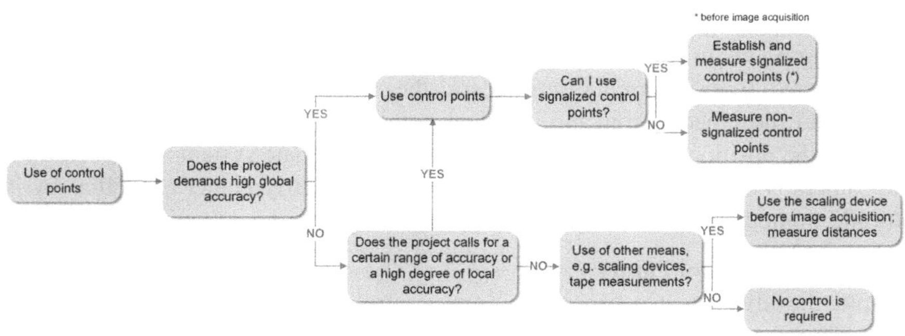

Fig. 6.20 Decision support diagram for using control points in photogrammetric workflows

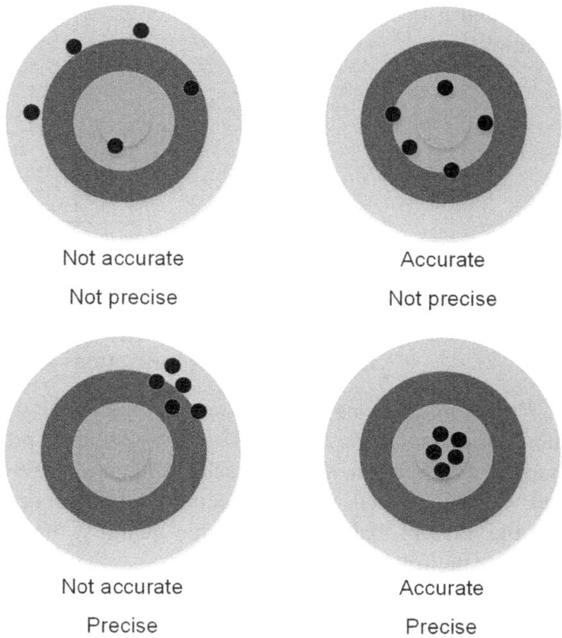

Fig. 6.21 Accuracy versus precision

can make estimates of the error for a measurement on a quantitative base. Since the error can be estimated, the accuracy of a measurement can also be estimated.

In general, errors are classified as systematic, which can be determined, and random, which are undetermined. The definitions of error, systematic and random are as following:

- Error, in general, is the outcome of a measurement minus a true value of the measurand.
- Systematic error is the mean that would result from an infinite number of measurements of the identical measurand that are executed repeatedly, minus a true value of the measurand.
- Random error is the result of a measurement minus the mean that would result from an infinite number of measurements of the identical measurand that are executed repeatedly as well.

A systematic error is caused by a fault in the method used or by an incorrectly operating instrument or by the user. If the procedure suffers from a systematic error, then the mean value will be always different from the true value. Random errors are not able to be avoided. They are inescapable because any measurement has a limiting rule which reflects to an uncertainty.

Usually, it is described by the Root Mean Square Error (RMS). Regarding the terms used, ε is the error, i.e. the difference between the measurement and the true value, and n is the number of measurements. The RMS is given by the following equation:

$$RMS = \pm\sqrt{\frac{\sum_{i=1}^{n} \varepsilon_1^2}{n}}$$
$$RMS = \pm\sqrt{\frac{\sum_{i=1}^{n} (\hat{y}_i - y_i)^2}{n}} \qquad (6.20)$$

$y_1, y_2, \ldots, y_n$ are the observed values of the sample items
$\hat{y}_i$ is the true value of those observations
n is the number of observations in the sample.

Precision or geometric precision expresses the distribution of a set of measurements about an average value. Usually, geometric precision is described by another statistical expression, namely standard deviation. If n is the number of measurements and v is the residual, which reflects to the difference between a measured quantity and the most probable value for that quantity, the precision of a sample dataset is given by the following equation:

$$\sigma = \pm\sqrt{\frac{\sum_{i=1}^{n} v_1^2}{n-1}}$$
$$\sigma = \pm\sqrt{\frac{\sum_{i=1}^{n} (x_i - \bar{x})^2}{n-1}} \qquad (6.21)$$

$x_1, x_2, \ldots, x_n$ are the observed values of the sample items
$\bar{x}$ is the mean value of those observations
n is the number of observations in the sample.

Control points are clearly identifiable points in the object space whose locations, and thus coordinates, are known. In practice, they are used to verify positioning of close-range or aerial images, remote sensing images or map features. If the points are located on the ground, they are called Ground Control Points (GCPs).

Even though they are usually called control points, sometimes a number of these points are used as checkpoints. They are commonly called checkpoints because their use is dedicated to perform an independent, quantitative evaluation of image location error. In fact, they are acting as validation points.

In photogrammetry, as the control points can be used in image orientation processes, some of these points are not used in the initial processes, and they are considered and used as checkpoints. In practice, both they are established in the building or around the building; they are measured the same way, but they are grouped and used in different photogrammetric processes. Regardless of whether the target's being used as control points or checkpoints, the characteristics and criteria for collection are the same. The difference between control points and checkpoints is only in their

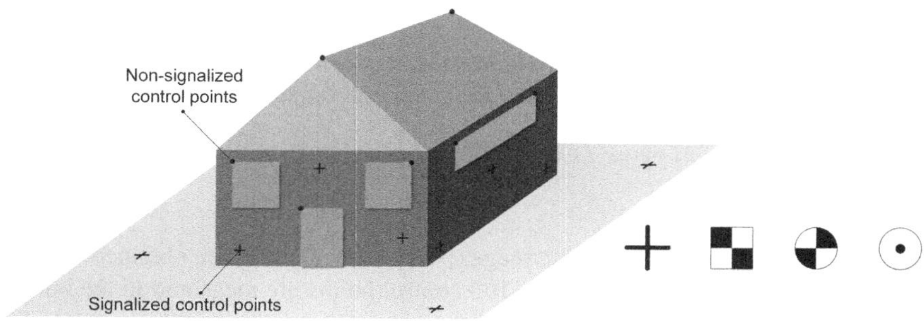

Fig. 6.22 The control points' types. Different types for signalized control points are given on the right

use, not in their collection workflow. Control points are used in the image orientation process while checkpoints are used only in the evaluation check.

Control points and checkpoints can be either signalized or non-signalized, depending the case each time. Signalized points means that specific target types are used to locate, indicate and facilitate their measurement. It is not always easy to use signalized points for many and different reasons. For instance, it is not permitted to stick such targets on a historic building, or it is impossible to reach every part of the building. Commonly, in a historic building, non-signalized points are used. These are "natural" points which are chosen by the users in order to complete the photogrammetric network points. The points are key features on the object, and they should be clearly and easily located, be visible, measured without difficulty and used during the photogrammetric processes. A sketch should be always used to clearly outline and depict the locations of non-signalized points. An indicative example from setting up a network of such points is given in Fig. 6.22.

The size and the features of the signalized points should be carefully designed and used. The targets, such as the ones presented in Fig. 6.22, must be properly designed using the appropriate line type, outline and color, while their size it is of great importance so as to be clearly and precisely identifiable in the image during the photogrammetric processes.

Exterior Orientation of a Single Image

Exterior orientation is the way in which the image and the object space are connected. It is accomplished by determining the camera position and the orientation parameters in the space, i.e. in the object coordinate system. The camera position is determined by the location of its perspective center and by its attitude, which is expressed by three rotation angles, i.e. ω, φ and κ, as illustrated in Fig. 6.9.

The exterior orientation of a camera, i.e. the position and orientation of the camera with respect to the object coordinate system, can be determined by using the collinearity equations, as previously discussed (Sect. 6.4.2). The collinearity equations (6.3) express measured quantities, i.e. the coordinates (x, y), of a point in the image plane, as a function of the exterior orientation parameters. As a result, the collinearity equations can be directly used as observation equations. The following functional Eq. (6.22) illustrates the relationship between the point image coordinates, the exterior orientation parameters and the point coordinates in the object space.

$$x, y = f(\underbrace{X_o, Y_o, Z_o, \omega, \varphi, \kappa}_{\text{exterior orientation}}, \underbrace{X, Y, Z}_{\text{object point}}) \tag{6.22}$$

For every measured point in the image space, two equations are obtained, one for the abscissa x and one for ordinate y. If 3 control points are measured in the image, and their coordinates are also known in the object coordinate system, a total of 6 equations are formed. The solution of this system leads to the 6 parameters of the exterior orientation. Having more than 3 control points, more than enough equations are formed and thus least squares adjustment should be performed for determining the unknown parameters. The collinearity equations are not linear in the parameters and must be linearized with respect to these parameters. It also requires approximate values of the unknown parameters, with which the iterative process will start.

Various methods have been developed for single image orientation, which are based on the geometric and topological characteristics of imaged objects. In their study, Grussenmeyer and Khalil (2002) present a survey of classical and modern photogrammetric methods for the determination of the exterior orientation parameters, some of which are available as software packages.

Taking the advantage of a free software for academic use, namely PhoX (Luhmann 2018) developed by Prof. Dr.-Ing. Thomas Luhmann at the Institute for Applied Photogrammetry and Geoinformatics, Jade University of Applied Sciences, a typical photogrammetric resection is provided as an example.

Having as input an image of a building (Fig. 6.23), the interior orientation of the camera, the object coordinates of 6 control points, and the image measurements of the 6 control points, the determination of the exterior orientation parameters is performed based on the collinearity equations. This photogrammetric space resection method provides a non-linear solution that requires a minimum of 3 control points and approximate values for the 6 unknown exterior orientation parameters. The input data for this example is provided by Luhmann (2018).

The calibration parameters of the camera used for this example are given in Table 6.3, while the measured image and the control points coordinates are given in Table 6.4. The output of this photogrammetric resection example, i.e. the 6 parameters of the exterior orientation, $X_o, Y_o, Z_o, \omega, \varphi, \kappa$, is given in Table 6.5.

6.5 Stereo and Multi-image Photogrammetry

As images are only two-dimensional products and can be expressed by a *2D* plane, the location of any point in an image can be described with just two coordinates, i.e. (x, y). However, as the real world is three-dimensional, the location of any point

Fig. 6.23 Photogrammetric resection for determine the exterior orientation parameters

Table 6.3 Interior orientation parameters of the camera used in the photogrammetric resection example (*Source* Luhmann 2018)

Parameters	Calibrated values
$c(mm)$	23.959000
$x_o(mm)$	0.058000
$y_o(mm)$	−0.123700
$K_1(mm^{-2})$	−0.092689
$K_2(mm^{-4})$	0.082158

Table 6.4 Data for photogrammetric image resection; image and control points coordinates (*Source* Luhmann 2018)

Point no.	$x(mm)$	$y(mm)$	$X(m)$	$Y(m)$	$Z(m)$
10040	−8.4000	−5.1967	23.6487	−3.8535	11.4041
10039	−4.4580	2.3208	23.7468	−6.3577	16.2011
10038	1.8888	1.1842	23.8347	−12.1641	16.1227
10037	3.4712	−3.5589	23.8161	−13.7143	12.2731
10047	6.9136	−3.5496	21.9394	−15.8084	12.2037
10048	8.2729	2.0227	20.6042	−16.3718	16.7023

Table 6.5 Photogrammetric image resection results; the exterior orientation parameters

Image no.	X_o(m)	Y_o(m)	Z_o(m)	ω(°)	φ(°)	κ(°)
1	9.76735	0.93611	11.85151	74.31815	−129.34145	−13.79650

in the object space can be described by three coordinates, e.g. expressed by the *3D* cartesian coordinates (*X, Y, Z*).

As already stated, photogrammetry is the "*the art, science, and technology of obtaining reliable information from nonconduct imaging and other sensor systems about the Earth and its environment, and other physical objects and processes through recording, measuring, analyzing and representation*" (ISPRS 2018). In other words, by using *2D* images and performing *2D* measurements in the images, photogrammetry provides all the methods and techniques to determine *3D* coordinates in the real space. In order to do that, the information that was lost while capturing the image, i.e. from *3D* to *2D*, needs to be recovered.

Let's take a look at the following figure (Fig. 6.24). The light that strikes a given pixel in the image (point *p*) could have come from any point (e.g. P', P'', P''') along the ray from the pixel, through the perspective center, into the scene. This is obvious from the collinearity principle, which indicates that a point in the object space, the perspective center and the point in the image lie across a straight line.

Is there a solution in this problem? Yes, indeed, there is a solution. Adding another image acquired from a different location allows the intersection of two rays and this way to determine the *3D* location of the point where the light came from. The more the images, the better for the redundancy of the system, and thus of the determination of the point in the object space. As illustrated in Fig. 6.25, the light that strikes a given pixel in the left and right image (point p_1 and p_2) could come only from one point, i.e. point *P*, which is the intersection of the 2 rays, through the perspective center into the scene.

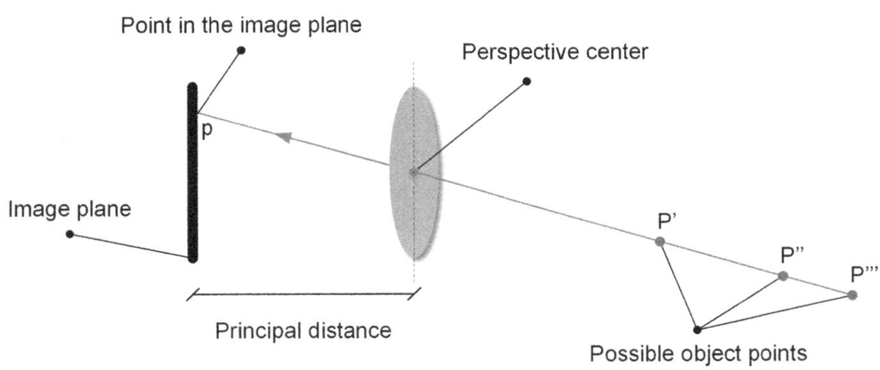

Fig. 6.24 Collinearity principle: one ray and the uncertainty in determining a point in the object space

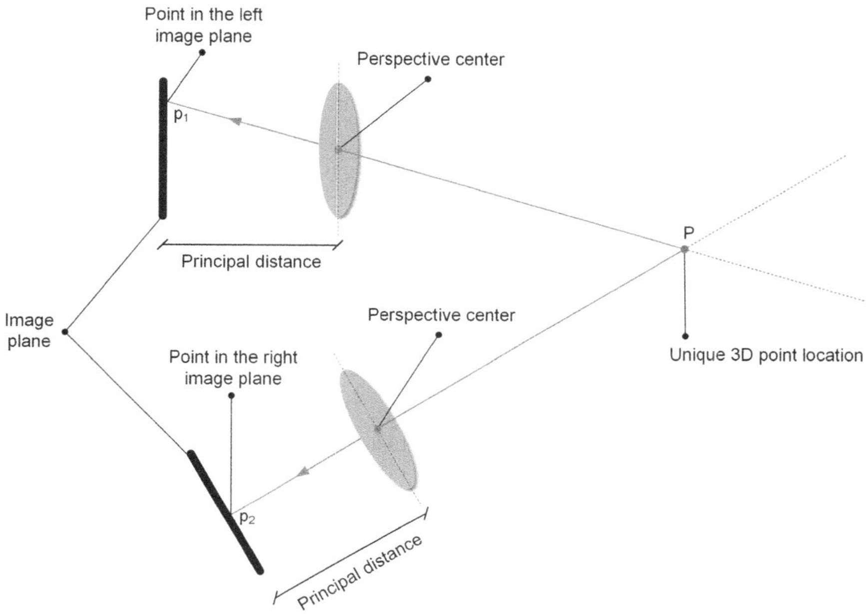

Fig. 6.25 Collinearity principle: two rays and the determination of a point in the object space

The essence of photogrammetry is to measure points in the image space and calculate the *3D* coordinates of these points in the object space. Based on the structure of Fig. 6.25 and the collinearity equations as already expressed (6.3), the following information is needed in order to determine the (unique) *3D* point location:

1. The interior orientation of the camera used, as the measurements will take place in the image plane. The (calibrated) values of x_o, y_o, c are a prerequisite, while the lens distortion parameters if needed.
2. The exterior orientation of the camera used, i.e. the perspective center position for each camera used as well as the orientation (rotation) of each camera about its perspective center. Usually, expressed via the parameters X_o, Y_o, Z_o, ω, φ, κ.
3. The location of the point P on each image sensor, i.e. p_1, p_2, etc., which is expressed by the image coordinates (x, y) in the image plane.

6.5.1 Interior Orientation

What has been said so far about photogrammetric methods has been the assumption that images are captured by metric or non-metric cameras. In the case of digital images instead of known analogue images, the image coordinates of the pixels have a different reference system than what has been seen so far. Since the *i* and *j* pixels

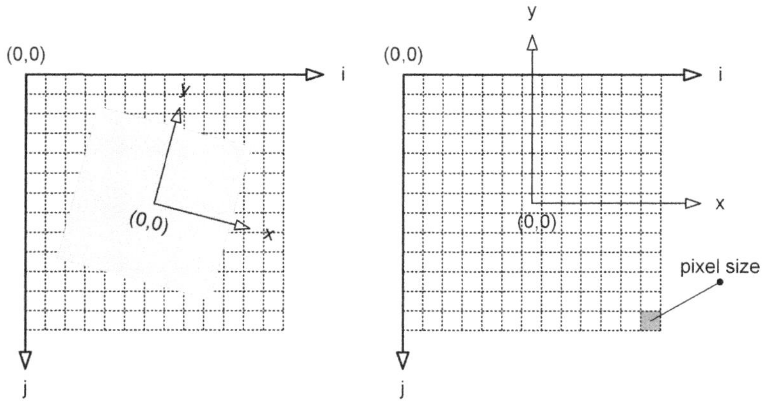

Fig. 6.26 The relationship between pixel-based coordinates and image coordinates

of the digital images are not referenced in the principal point but in their upper left corner, the collinearity condition and the relevant equations will provide wrong results.

In order to restore the central projection model and thus to allow the photogrammetric methods described so far to be applied, i and j pixel coordinates should be transformed into a x, y coordinate system that is centered to the principal point. The most common way to do that is to use an affine transformation (6.23), as shown in Fig. 6.26 on the left.

$$\begin{aligned} x &= a_1 + a_2 \cdot j + a_3 \cdot i \\ y &= a_4 + a_5 \cdot j + a_6 \cdot i \end{aligned} \tag{6.23}$$

where

i, j	are the pixel-based image coordinates
x, y	are the image plane coordinates
$a_1, a_2, a_3, a_4, a_5, a_6$	are the coefficients of the affine transformation.

Essentially, affine transformation is a rotation, a translation, and a scale, with 6 unknown parameters, that is why it is called a 6 degrees of freedom (DOF) representation. For the calculation of the unknown parameters, 3 common points with known coordinates should be available in both systems. Typically, more than 3 points are available, and a least squares adjustment is performed. In the case of the analogue image, the fiducial marks are used.

However, many times, like in the case of digital images where fiducial marks are not available, instead of the affine transformation, for the transformation of the image pixels to the image coordinate system, the following equations are used (6.24), as shown in Fig. 6.26 on the right.

$$x = (i - i_c) \cdot ps_i$$
$$y = (j - j_c) \cdot ps_j \qquad (6.24)$$

where

i, j are the pixel-based image coordinates
x, y are the image plane coordinates
i_c, j_c are the pixel-based image coordinates at the center of the image
ps_i, ps_j are the pixel size in i and j respectively.

The above Eqs. (6.24) are usually used when digital images are not originating from digitized photographs. This is the case of a typical digital camera which is not considered as a metric camera. As said, in this case, there are no fiducial marks available, and apart from the coordinate system transformation for the interior orientation, it is necessary to follow the calibration protocol to determine the calibrated values of the camera internal geometry.

6.5.2 Relative Orientation

Relative orientation is the process in which two overlapping images of a stereo pair are related to each other in an arbitrary coordinate system so that their relationship is the same as it existed at the time of image acquisition. At this first stage, after the interior orientation, all images must be "connected".

The aim of relative orientation is to restore the same (central) projection conditions between the stereo pair, as they incurred at the time of image acquisition, and it is depicted in Fig. 6.27.

Besides, all these occur as the ultimate goal remains on how to determine points in the object space from two or more images. Consider the example of the two images illustrated in Fig. 6.28, where the two images (or in general the images) should be taken from different points.

A drone is used to capture two consecutive images of a building from an aerial view. The result is two different perspective views of central projection. The displacements due to the elevations (relief) are different in the two images. Look at the back side of the building in the images. The same is for the building's left side. In photogrammetry, the difference of point position in two images is called *parallax*. Different deflections in the two images due to fluctuations of the relief create parallax differences. As a result, the elevation differences in an object, e.g. a building, reflect to parallax differences in image pairs. Taking the advantage of these differences, it is possible to measure the elevation or the height of a building. In any case, it is just the central projection, i.e. the variable scale and aberrations due to changes in the relief, that allow the determination of a point in the 3D space from two images.

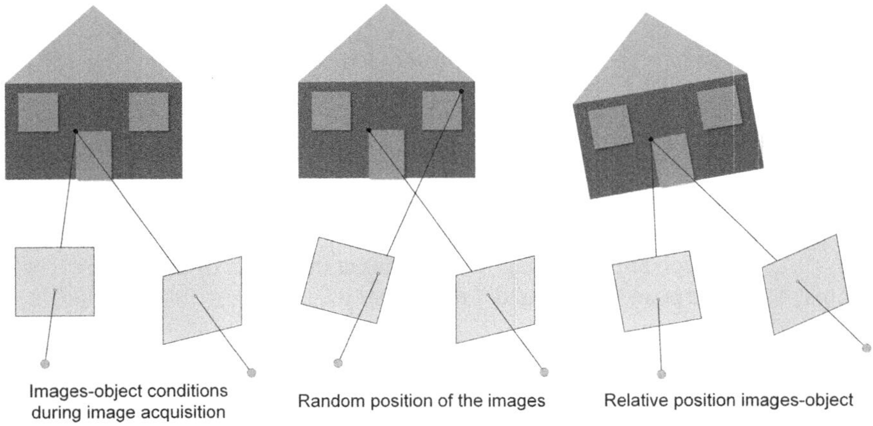

Fig. 6.27 Graphical representation of relative orientation principle

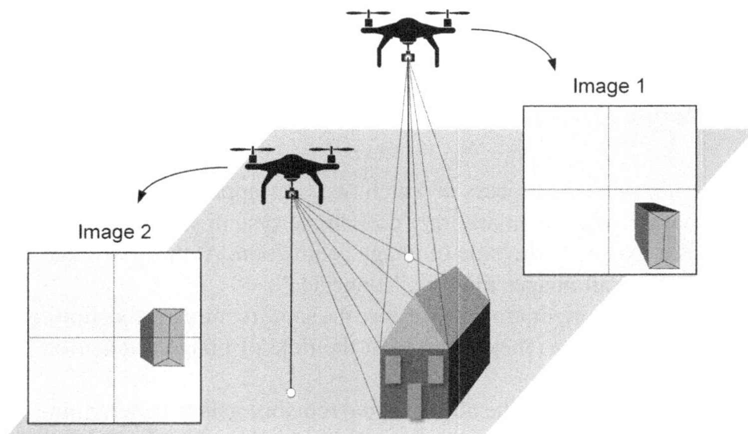

Fig. 6.28 The displacements due to elevation and the parallax

As it is shown in Fig. 6.29, parallax along *x-axis*, it is the x component of the deviation which is caused due to elevation. On the other hand, *y-parallax* component is only affected by the course of the image acquisition mean, e.g. the drone. This way, *x-parallax*, i.e. the displacement $A - A'$ in Fig. 6.29, is the algebraic difference of the image coordinates along the *x-axis*, given by formula (6.25). Essentially, it is the apparent movement of the ground point on the image caused by the camera motion. The higher the point elevation, i.e. the closest to the camera, the higher the value of the *x-parallax*.

$$p_x = x - x^{'} \tag{6.25}$$

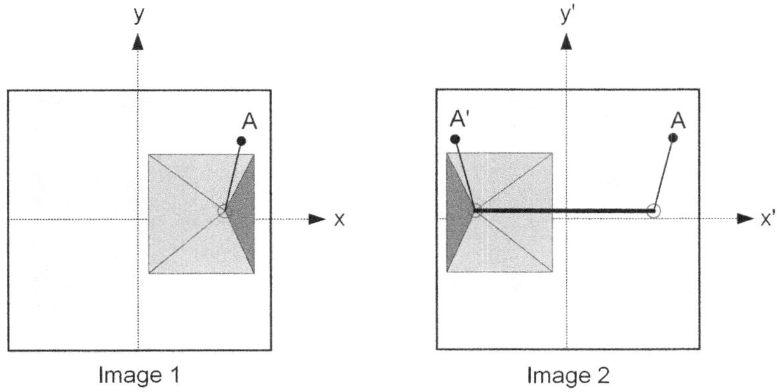

Fig. 6.29 Definition of parallax

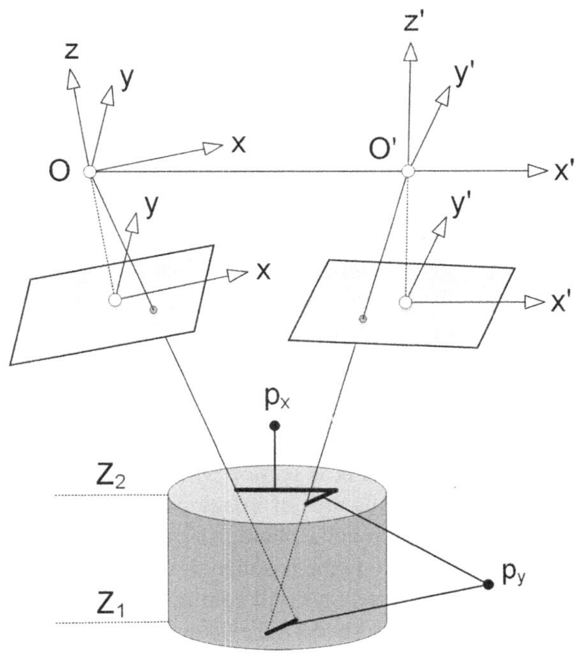

Fig. 6.30 Parallax components and image stereo pair coordinate system for relative orientation

Better understanding for the role of parallax and its effect in relative orientation and stereo viewing is possible through understanding the notion of the following figure (Fig. 6.30). From this figure, two very important judgements are extracted:

- *x-parallax* is connected with the elevations/heights as its magnitude varies from Z_1 to Z_2.
- *y-parallax* is connected with stereo viewing as its elimination secures the intersection of the homologous rays and makes certain that stereo viewing is available.

Table 6.6 Interior orientation parameters of the camera used in the relative orientation example (*Source* Luhmann 2018)

Parameters	Calibrated values
$c(mm)$	47.423900
$x_o(mm)$	0.050600
$y_o(mm)$	0.212100
$K_1(mm^{-2})$	−1.068300000E-0005
$K_2(mm^{-4})$	2.669700000E-0009

In practice, relative orientation depicts the status of an image stereo pair, i.e. the translation and rotation of one image (e.g. the left) with respect to the other (e.g. the right) in a local *3D* model coordinate system which is utilized to bind this model. As it is shown in Fig. 6.30, this *xyz* local *3D* model coordinate system is located in the perspective center of the left image, while it is oriented parallel to its image coordinate system. In this case, the exterior orientation parameters of the left image with respect to the model coordinate system can be considered as $x_L = y_L = z_L = \omega_L = \varphi_L = \kappa_L = 0$. Therefore, if the left image is considered as "fixed", then the right image is oriented in the same model coordinate system by using 3 translations and 3 rotations parameters which are expressed by:

$x_R = b_x$ the x component of baseline b
$y_R = b_y$ the y component of baseline b
$z_R = b_z$ the z component of baseline b
ω_R the rotation angle around x-*axis*
φ_R the rotation angle around y-*axis*
κ_R the rotation angle around z-*axis*.

Baseline b between the 2 perspective centers O and O' is split in three components, i.e. b_x, b_y, b_z. In the next section where epipolar geometry is discussed, it is shown that the homologous rays of image point pairs belong on the same plane with the baseline, i.e. the epipolar plane. Assuming that the right image and its perspective center is moved along the baseline towards O and also that the image is not rotated, then the homologous rays will continue to be coplanar with b. The careful study of the similar triangles shows the evident principle; the scale of the image stereo model is directly proportional to b. Practically, this means that the model coordinate system can be scaled randomly, relying on b. As a result, one of the 3 components of b is fixed to a constant value, and this usually happens by assigning $b_x = 1$. Consequently, five parameters are available for determination, i.e. $b_y, b_z, \omega_R, \varphi_R, \kappa_R$, and the definition of relative orientation.

By using the PhoX software (Luhmann 2018) and having as input a stereo image pair (Fig. 6.31), the interior orientation of the camera (Table 6.6), the image measurements of the 6 points, the determination of the relative orientation parameters is achieved. The input data for this example is provided by Luhmann (2018).

Fig. 6.31 Image stereo pair (left and right image) and the points' measurements to determine relative orientation parameters

Table 6.7 Measurements on the left and right image for relevant orientation determination

Point no.	$x_L(mm)$	$y_L(mm)$	$x_R(mm)$	$y_R(mm)$	$p_y(mm)$
1	−0.5202	14.0109	−18.2356	10.0988	0.0000
2	0.1500	2.2904	−17.2530	−1.4864	−0.0001
3	2.0792	−11.8254	−14.7155	−15.9415	0.0000
4	21.5280	15.8114	3.7876	12.3202	−0.0000
5	21.2224	2.9252	4.3227	−0.3414	0.0001
6	21.6270	−11.5941	5.3130	−15.1859	−0.0000

Table 6.8 Relative orientation results

	b_x(m)	b_y(m)	b_z(m)	$\omega(°)$	$\varphi(°)$	$\kappa(°)$
	1.00000	0.00416	−0.01755	4.00232	−0.40735	−1.33663
St. deviation	–	0.00008	0.00001	0.00161	0.00117	0.00027

The following 2 tables are illustrating the measurements on the left and the right image (Table 6.7) and the results of the relative orientation (Table 6.8). The mean *y-parallax* was calculated almost zero as it is shown in the following tables. This means that relative orientation has been restored properly. Now, under this condition, it is possible to generate the epipolar images and introduce stereo viewing, as the *y-parallax* has been eliminated.

The success of relative orientation can be also checked by supplementary point measurements in the model. As shown in Fig. 6.32, by measuring a new point (100) on the left image, the epipolar line on the right image is helping the user to pick the measurement of the homologous point in the right image. The direction of the epipolar lines is passing exactly over the candidate homologous point.

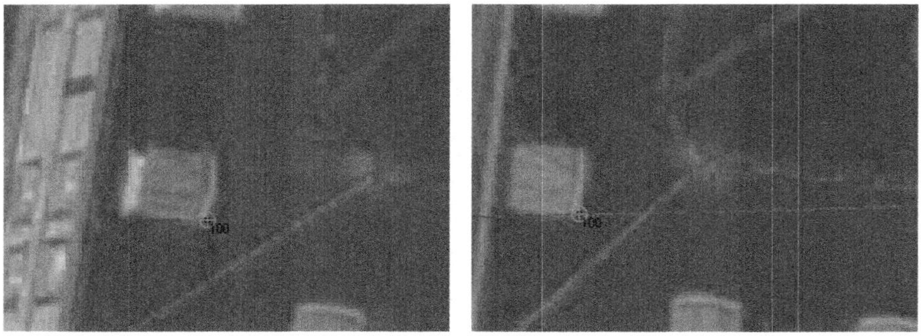

Fig. 6.32 Point measurement in the model after the determination of relative orientation parameters. The position of the epipolar line over the homologous point

Fig. 6.33 Epipolar geometry in an image stereo pair

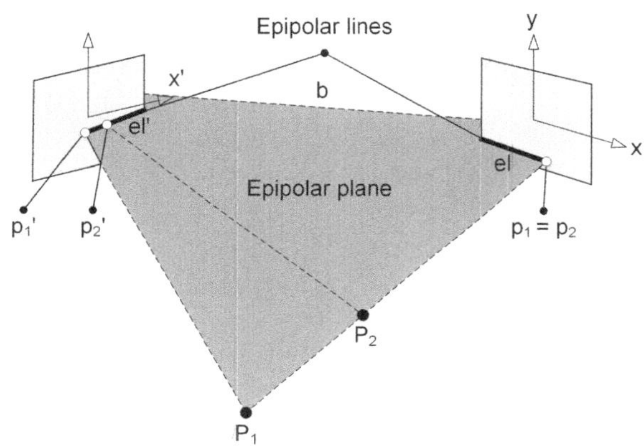

Epipolar Geometry

The geometry of an image stereo pair is illustrated in Fig. 6.33. An object point P_1 is observed in the two images, i.e. p_1 on the left and $p_1^{'}$ on the right image. Baseline b which connects the two perspective centers O and $O^{'}$ of the left and the right image respectively (not visible in this figure), and the two rays towards point P_1, i.e. $Op_1 P_1$ from the left and $O' p_1^{'} P_1$ from the right image, define a plane. This plane is called *epipolar plane* and intersects the two image planes along the lines el and el' respectively. These lines are called *epipolar lines*.

Epipolar geometry is a very important tool for photogrammetry. Considering an error-free ray intersection from the left and the right image, an object P_1 which reflects to an image p_1 in the left image, the corresponding point $p_1^{'}$ in the right image, must definitely lie on the epipolar plane ep and hence without doubt along the epipolar line el'. Epipolar geometry is momentous for various photogrammetric process as for example:

- When searching for conjugate points in image matching, the searching space is significantly reduced along the epipolar line.

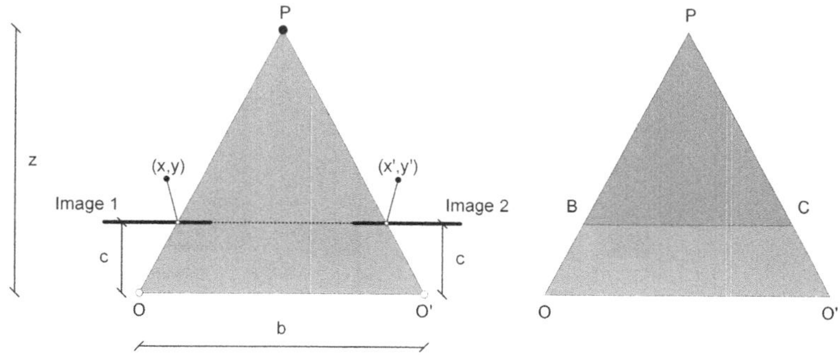

Fig. 6.34 Disparity in stereo images

- Use of epipolar geometry to generate normalized images -also called epipolar images- and thus restrict the search of conjugate points along the same scan lines.
- Assuming a new object point P_2 along the ray Op_1P_1, it is easily perceived that the depth difference between P_2 and P_1 results in a parallax along the epipolar line el'.

According to Cho et al. (1992) most algorithms in computer vision and digital photogrammetry suppose to be the case that digital image stereo pairs are registered in epipolar geometry (normalized images). The reason is to reduce the searching space for the conjugate points along the same scan lines. In their study, they describe the procedure of generating normalized images of aerial photographs with respect to the object space. In a similar work, Tsioukas et al. (2000) deal with the same issue, but in that study for generating epipolar images in close range images.

Disparity Map

Disparity is the term used more in computer vision, and it is one of the most important aspects of stereo viewing. It concerns the displacement between two consecutive images. In order to better understand the disparity concept and the determination of depth, i.e. height/elevation, let's consider Fig. 6.34.

The two cameras are connected as shown in Fig. 6.34 so as their optical axes are parallel while their baseline, i.e. the distance between the lens center, is indicated by b. Suppose that this line is perpendicular to the optical axes and set the *x-axis* to be parallel to the baseline. The coordinates of point $P(X, Y, Z)$ are measured relative to the center of the axes located midway between the two lenses. If c is the principal distance of the camera used, the coordinates of point A in the left and right image are (x, y) and (x', y') respectively, then from the equivalent triangles POO' and PBC, the following equations can be extracted:

Fig. 6.35 Image stereo pair and the disparity map

$$\begin{aligned} \frac{b}{x - x'} &= \frac{Z}{c} \\ \frac{b}{p_x} &= \frac{Z}{c} \\ Z &= \frac{b \cdot c}{p_x} \end{aligned} \tag{6.26}$$

From the Eqs. (6.26), it becomes evident that the coordinates of any point, like point P, are inversely proportional to the displacement, resulting in close-range photogrammetry objects that these coordinates can be accurately measured, which is not the case for distant objects. In addition to that, the distance between the two cameras, i.e. baseline b, is directly proportional to the displacement. Thereby, having a given error in determining the displacement, the accuracy of depth determination increases with the increase of the baseline. The displacement is also proportional to the principal distance, since the images are magnified as the principal distance increases.

Figure 6.35 provides an expressive example of 2 stereo images which were processed to produce the disparity map. The input images, which have been retrieved from MiddleburyCollege (2015), are rectified such that the corresponding points are located on the same rows.

6.5.3 Absolute Orientation

Absolute orientation means that all data are registered in the real world at the object coordinate system by using known points, either control points or the camera stations. In order to do so, at least three known points are required in projection, regardless the number of the images used. In fact, it describes the transformation of the local model coordinate system *xyz*, as previously described in relative orientation Sect. 6.5.2. The transformation provides the transition from a position, rotation and scale based on random choice into the real world through the object coordinate system *XYZ*

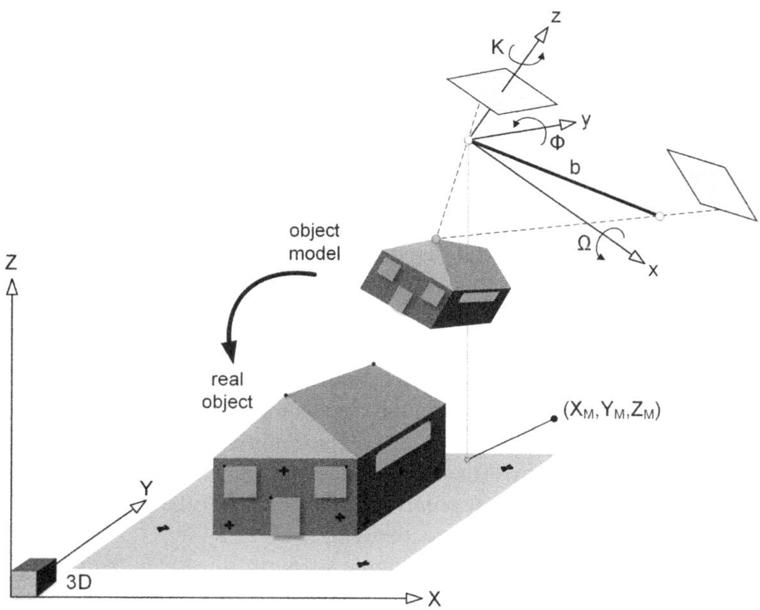

Fig. 6.36 Setup of absolute orientation

(Fig. 6.36). Mainly, this is achieved through the use of control points. As already discussed in Sect. 6.4.5, control points are object points measured in the model coordinate system. The control points are possible to have one or more known coordinate features in object space, e.g. *XYZ*, *XY* only or *Z* only.

A similarity transformation with 3 translations, 3 rotations and one scaling factor restores the absolute orientation. The relation between the model coordinates *xyz* and the object coordinates *XYZ* can be expressed by the following Eqs. (6.27):

$$\begin{bmatrix} X \\ Y \\ Z \end{bmatrix} = \begin{bmatrix} X_M \\ Y_M \\ Z_M \end{bmatrix} + m \cdot R_{\Omega\Phi K} \cdot \begin{bmatrix} x \\ y \\ z \end{bmatrix} \tag{6.27}$$

where

X_M, Y_M, Z_M	are the object coordinates of the *xyz* model system origin
m	is the scale factor of the *xyz* model system
R	is the *3D* rotation matrix of the *xyz* model system into the *XYZ* object coordinate system
X_o, Y_o, Z_o	is the camera projection center coordinates in the object coordinate system.

The 7 unknown parameters of the absolute orientation $X_M, Y_M, Z_M, \Omega, \Phi, K, m$, can be determined through at least seven equations. According to Eq. (6.27) it is possible to formulate:

- 3 equations for a full control point where *X, Y, Z* are all known.
- 2 equations for a horizontal control point where only *X, Y* are known.
- 1 equation for a height control point, where only *Z* is known.

As a result, the absolute orientation requires at least 2 horizontal control points and 3 height control points (non-collinear horizontally). Another option is that absolute orientation requires 2 full control points and a height control point not collinear horizontally with the full control points.

Continuing from the same data set example presented in relative orientation (Sect. 6.5.2), the absolute orientation of the stereo model is performed. The appropriate measurements of 4 (full) control points are accomplished properly. The distribution of all points, which is across the stereo model coverage, is illustrated in Fig. 6.37. The measurements of the control points and the results of the absolute orientation are given in Tables 6.9 and 6.10 respectively.

The last figure (Fig. 6.38) in this example scenario illustrates the quality of the absolute orientation results. As depicted in this figure, the differences between the control points' measurements and the projections of the control points, considering

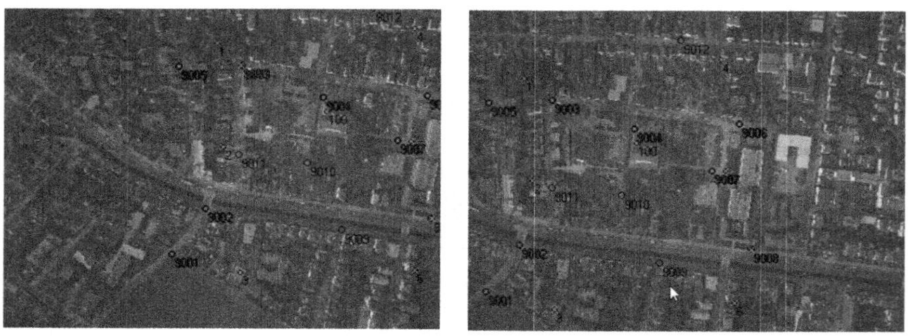

Fig. 6.37 Points measurement for absolute orientation parameters determination

Table 6.9 Stereo model and object coordinates of the 4 control points used in the absolute orientation of the image stereo pair

Point no.	$x(m)$	$y(m)$	$z(m)$	$X(m)$	$Y(m)$	$Z(m)$
9001	−341.400	−602.900	−2917.400	446287.948	5888257.964	3.560
9005	−289.600	672.700	−2811.900	446285.857	5888522.059	4.233
9006	1378.400	495.400	−2854.600	446631.083	5888501.479	4.524
9008	1447.800	−353.200	−2926.500	446653.816	5888326.842	3.809
9010	576.100	28.400	−2881.600	446470.651	5888397.207	3.697
RMS	*0.061*	*0.022*	*0.078*	–	–	–

Table 6.10 The absolute orientation results

$X_o(m)$	$Y_o(m)$	$Z_o(m)$	$\Omega(°)$	$\Phi(°)$	$K(°)$	m
446339.667	5888432.720	594.184	−4.5422	−1.2118	2.6720	206.1567

Fig. 6.38 The displacements of the control points; measured position and position based on the object coordinates

their true coordinates and using the absolute orientation parameters, are extremely small. This picture shows what the statistical elements of the solution demonstrate as well, confirming the relative orientation accuracy.

6.5.4 Bundle Adjustment

The photogrammetric triangulation, more usually known as bundle block adjustment or simply bundle adjustment, it is a very important process in close-range photogrammetry. The bundles of rays of the images are tied with common object points. Hence, it is a method that connects small or huge number of images by using the bundles of rays that associate points in the image and object space. From the practical point of view, control points and tie points are used to merge this model of all images with an object coordinate system.

A tie point is a distinctive feature that can be identified with clarity in two or more images and that can be selected as a reference point. Tie points do not have known object (i.e. ground) coordinates, but they can be used to extend control over areas where there are no control points. They are used in order to identify how the images in a project relate to each other.

The more the bundle rays are connecting the image space with the object space, the better for the stability of the overall model. However, the most serious prerequisite which should be followed, in any case, is that the geometry of the bundles of rays must be almost ideal. This means that the homologous rays from the relevant images should converge in the "same" location, i.e. in almost the same position in the object space.

Since the bundles of rays are much more than the needed ones, these reflects to an over-determined system of (collinearity) equations, an adjustment technique is utilized. The outcomes of the bundle adjustment are the:

- *3D* coordinates of the points in the object coordinate system.
- Interior orientation parameters for all images used, in the case the cameras are calibrated simultaneously.
- Exterior orientation parameters for all images used.
- Statistical parameters which are necessary to analyze the bundle adjustment quality (image coordinate residuals, standard deviations of object points, standard deviations of orientation data, etc.).

In fact, bundle adjustment serves the simultaneous numerical fit of an unlimited number of images which are spatially distributed across the object that is under photogrammetric measurements. All observations and all unknown parameters of a photogrammetric measurement project are considered within one calculation package done at the same time. For this reason, the bundle triangulation is assumed as the most robust and accurate method in photogrammetry with respect to image orientation and point determination.

In close-range photogrammetry, image configuration for bundle adjustment is not like the aerial image configuration where the use of regular strip arrangements of images is considered as the normal case. The close-range configuration looks like the one presented in Fig. 6.39. Usually, it is described by irregularly organized image configurations, where normally images are not captured with a metric camera.

The bundle adjustment workflow is based on the input data which are normally the image coordinates measured by manual or automatic photogrammetric techniques. This is able to lead to the *3D* reconstruction of the object, as represented by the observed and measured object points. But this is an arbitrary presentation of the

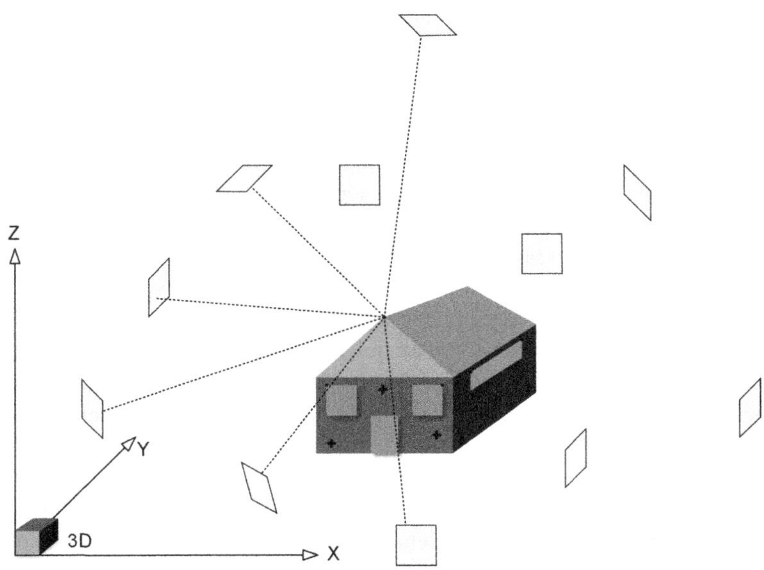

Fig. 6.39 Image configuration in close-range photogrammetry bundle adjustment

object, and for this reason, additional information in the object space (e.g. control points, distances, etc.) are introduced. This way, the scale, position and orientation of the model is defined with respect to the object coordinate system.

Bundle adjustment has a mathematical model that is based on the collinearity equations (6.16). These equations allow the direct formulation of the observed image coordinates as functions of all unknown parameters in the process. The collinearity equations are linearized at approximate values; however, in order to linearize the model, it is necessary to have approximate values. After that, they can be used directly as observation equations for a least-squares adjustment workflow based on the Gauss–Markov model (Luhmann et al. 2013).

As the images could be many in a bundle adjustment process, as well as the observed points, it is necessary to be sure that the appropriate number of observations are available to calculate all unknown parameters. An easy way to estimate the total number of available observations and all unknowns is the following:

- For each point, there are 2 observations (x, y) per image.
- Each unknown point in the *3D* object coordinate system has 3 unknown parameters (X, Y, Z).
- Each image has 6 unknown exterior orientation parameters ($X_o, Y_o, Z_o, \omega, \varphi, \kappa$).
- Interior orientation of each camera used, which reflects to *0* or $\geq$3 unknown parameters.

Assuming a typical example like the one illustrated in Fig. 6.39 for a building, the following table (Table 6.11) provides a clear picture for the number of observations and unknowns. In practice, this example gives the piece of information that indicates the necessary observations in order to perform a bundle adjustment.

Bundle Adjustment with DBAT Open Source Software

Damped Bundle Adjustment Toolbox (DBAT) is a series of functions written in MATLAB. The toolbox performs bundle adjustment calculations based on tie point observations provided by third party software. DBAT supports input from both EosSystem's Photomodeler and Agisoft PhotoScan. DBAT was originally conceived to test the influence of damping algorithms in bundle adjustment, which are applied to perform a more robust computation than the classical Gauss–Markov algorithm (Börlin and Grussenmeyer 2013a, b). As such, it supports several damping approaches, namely the Levenberg–Marquardt, Gauss–Newton with Armijo line search, and Levenberg–Marquardt with the Powell dogleg. It has since been used for other purposes, including robust camera calibration (Börlin and Grussenmeyer 2014) and

Table 6.11 Number of unknowns parameters in a close-range image configuration for bundle adjustment

Description	Number of ...	Number of unknowns	Total unknowns
Number of images	11	6	66
Number of points	20	3	60
Total number of unknowns			126

(a)

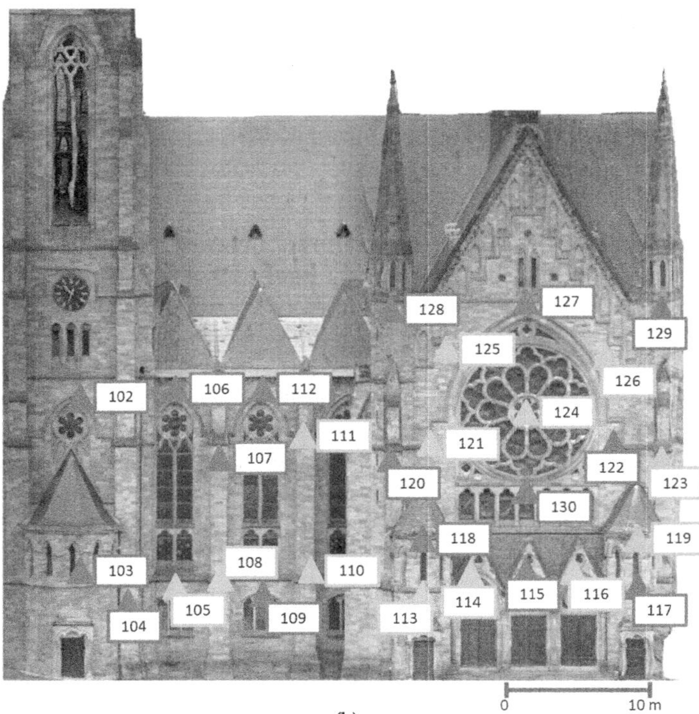

(b)

Fig. 6.40 The St-Paul church (Strasbourg, France) used in the case study (**a**). Sixteen control points (green triangle) and thirteen check points (orange triangles) scattered on the church's eastern façade (**b**)

Fig. 6.41 Visualization of the image orientation results in Agisoft PhotoScan (left) and DBAT (right)

bundle adjustment quality control (Börlin and Grussenmeyer 2016; Murtiyoso et al. 2017, 2018). DBAT has also seen a recent development towards modular bundle adjustment, which makes it simpler for users to investigate novel projection models.

A case study on the use of DBAT to reprocess an Agisoft PhotoScan project is described here. UAV images on the St-Paul church in Strasbourg, France are used. The purpose of the experiment was to be able to recreate Agisoft PhotoScan results under the same weighting conditions, while providing additional bundle adjustment metrics. These metrics are useful in terms of project quality control, as they enable the user to identify potential problems within the photogrammetric project which is otherwise undetectable in commercial software.

In this case study, only the eastern façade of the church is analyzed (Fig. 6.40). Sixteen control points and thirteen check points were scattered on this façade, which were measured using the spatial intersection method from two ground stations using a total station. A total of 485 images of this façade were captured using the DJI Phantom 4 Pro UAV. The images were processed using Agisoft PhotoScan and then reprocessed using DBAT in order to generate additional bundle adjustment metrics.

The results of the processing are illustrated in Fig. 6.41. Although Agisoft PhotoScan already furnishes a report for its projects, often times this report lacks details which are nevertheless useful for the (advance) user. DBAT not only recreates Agisoft PhotoScan results, but also generates diagnostics such as the standard deviations of

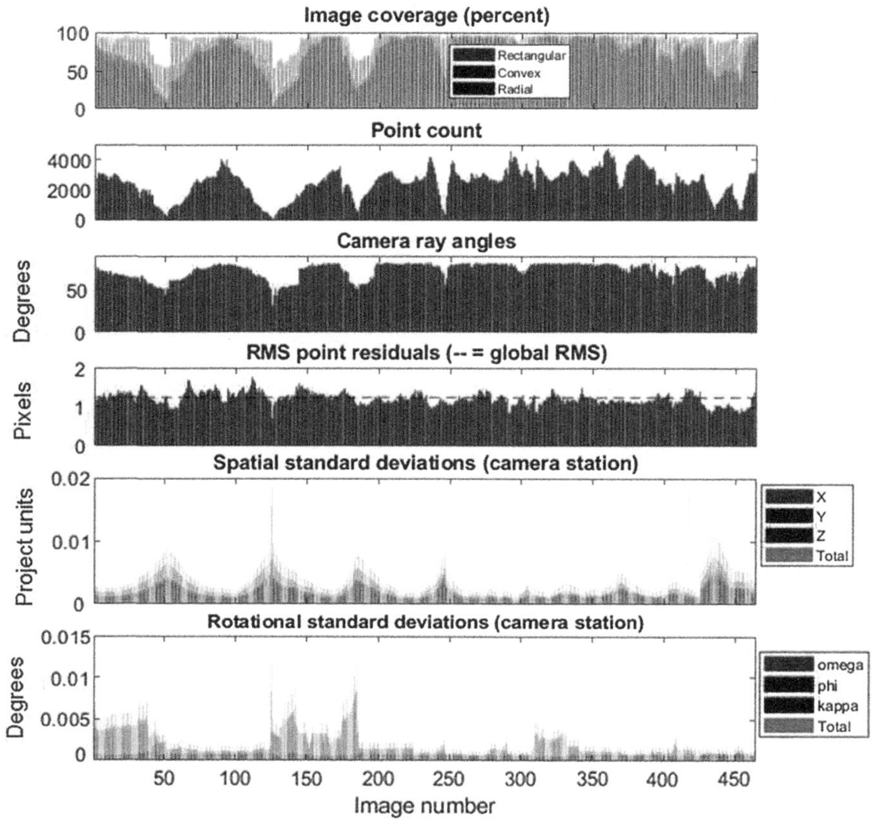

Fig. 6.42 Examples of statistics generated at the end of the bundle adjustment by DBAT

the exterior parameters, camera ray angles, correlation between interior parameters, etc. (Fig. 6.42) In this example, Fig. 6.42 shows that some spikes can be observed in the exterior parameter standard deviations. This can be an indication on which parts of the project may be modified in order to increase its overall precision. A detailed report file complements the graphs and may be useful to detect other problems with the network.

6.6 Structure from Motion

Structure from Motion (SfM) is an image-based modelling technique that has emerged from advances in computer vision and photogrammetry for estimating *3D* structures from *2D* image sequences. It can generate high fidelity, dense, *3D* point cloud of an object, like a historic building.

Images from many angles and distances can be used in a SfM workflow. It is not necessary to have prior knowledge of locations or pose. In general, SfM enables "unstructured" image acquisition from the ground (terrestrial, close-range) or from unmanned platforms such as drones.

The original idea is dated back to the late 70s when (Ullman 1979) examined the interpretation of SfM from a computational point of view. In fact, the question addressed is how the *3D* SfM objects can be inferred from the *2D* transformations of their projected images when no *3D* information is conveyed by the individual projections.

Since the 80s, SfM was transformed to a valuable tool for generating *3D* models from *2D* images. Compared to conventional photogrammetry, SfM uses algorithms to identify matching features in a set of overlapping images and determines camera position and orientation parameters from the differential positions of the multiple matched features. The algorithm that powers SfM for matching features is the Scale Invariant Feature Transform (SIFT). SIFT is so powerful, it allows corresponding features to be matched even with large variations in scale and viewpoint, even under conditions of partial occlusion and changing illumination (Lowe 1999).

By using as a basis these calculations, overlapping images can be used in order to reconstruct a sparse 3D point cloud model of the shot object. Normally, this extracted *3D* model from the SfM method is refined to a much higher quality resolution employing Multi-View Stereo (MVS) methods. In addition to that, a dense *3D* point cloud model can be produced. This reflects to a spatial density/resolution of *3D* points that, in some conditions, is comparable to that delivered from laser scanners.

Let's have a brief look at the high-level steps followed in SfM approach:

Step 1: The first step concerns the process of matching corresponding features and measure distances between them on the camera image plane d_1, d_2, as illustrated in Fig. 6.43a. As already mentioned, SIFT is key to matching corresponding features despite large variations in scale and viewpoint.

Step 2: Normally, when the matching locations of multiple points where two or more images are known, simply there is one mathematical solution for the images' point of acquisition Fig. 6.43a. For that reason, we are able to calculate:

- individual camera positions (X_1, Y_1, Z_1), (X_2, Y_2, Z_2)
- orientations i_1, i_2
- focal lengths f_1, f_2
- relative positions of corresponding features b, h

in only one and single step, which is well-known as bundle adjustment (Fig. 6.43b). That is where the expression SfM is originating, i.e. scene structure alludes to all the above mentioned parameters, while motion refers to the movement of the camera in different locations.

Step 3: Next, a dense *3D* point cloud and *3D* surface of the object is calculated using the:

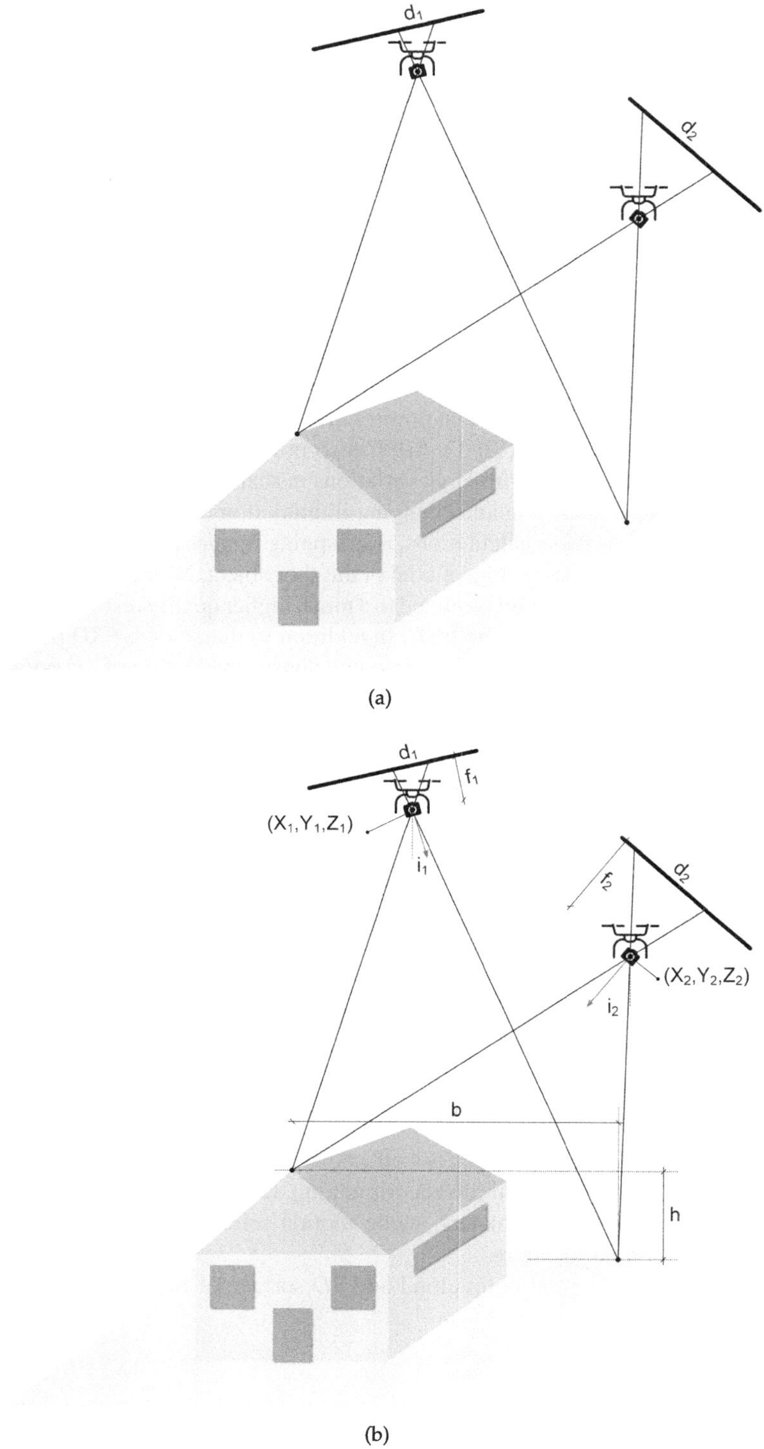

Fig. 6.43 Steps of SfM workflow; match corresponding features (**a**) and bundle adjustment (**b**)

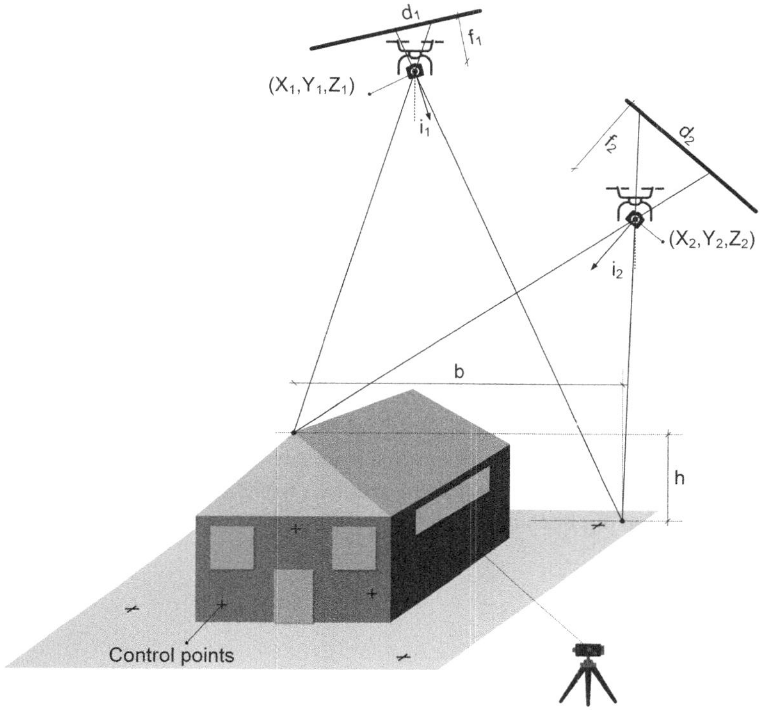

Fig. 6.44 Georectification in SfM workflow

- known camera parameters
- SfM points as "ground control" to reference the *3D* point cloud in an arbitrary coordinate system.

In practice, all pixels in all images are employed to determine the *3D* dense point cloud. This way, the dense model is almost identical in resolution to the raw images. Typically, it reflects in hundreds to thousands of points per square meter (*100* s – *1000* s *point*/m^2). This step is called MVS. A dense point cloud is a useful starting point for *3D* modeling that can be helpful in positioning a building into a *3D* scene.

Step 4: The next step concerns georectification. In general, it is a form of image rectification that transforms an image or a map into a common coordinate system. In this case, georectification reflects to the conversion of the *3D* point cloud from an internal and arbitrary coordinate system into a geodetic coordinate system. This can be achievable in one of two ways:

- Directly, having known the camera positions and focal lengths (e.g. from a previous step).

- Indirectly, by including a few control points with known coordinates across the object and around it. Normally, control points are measured either using conventional surveying techniques or GNSS measurements in order to determine their coordinates in the geodetic coordinate system (Fig. 6.44).

Step 5: A last step, which can be considered as optional, is the generation of derivative products. Such that the following could be considered:

- DSM: for instance, it can be expressed by a polygon mesh. In practice, it is a collection of edges, faces and vertices that define the shape of an object towards *3D* modeling. Usually, faces consist of triangles (triangle mesh), quadrilaterals, or other simple convex polygons.
- Orthoimage: for instance, it can be used for texture mapping. An orthoimage (or orthophoto or orthophotograph) is an image that was geometrically corrected ("orthorectified"). This way, the orthorectified image has a uniform scale. Different from an uncorrected image, an orthoimage can be used to measure true features, as it is an accurate representation of the object's real surface, having been adjusted for elevation, lens distortion, and camera tilt.

References

Albertz J (2001) Albrecht Meydenbauer - Pioneer of photogrammetric documentation of the cultural heritage. In: Proceedings of the 18th international symposium on CIPA 2001, Potsdam, Germany

Balleti C et al (2014) Calibration of action cameras for photogrammetric purposes. Sensors 14:17471–17490. https://doi.org/10.3390/s140917471

Börlin N, Grussenmeyer P (2013a) Bundle adjustment with and without damping. Photogramm Rec 28(144):396–415

Börlin N, Grussenmeyer P (2013b) Experiments with metadata-derived initial values and linesearch bundle adjustment in architectural photogrammetry. ISPRS Ann Photogramm Remote Sens Spat Inf Sci II-5/W1:43–48

Börlin N, Grussenmeyer P (2014) Camera calibration using the damped bundle adjustment toolbox. ISPRS Ann Photogramm Remote Sens Spat Inf Sci II-5:89–96

Börlin N, Grussenmeyer P (2016) External verification of the bundle adjustment in photogrammetric software using the damped bundle adjustment toolbox. ISPRS Arch Photogramm Remote Sens Spat Inf Sci XLI-B5:7–14

Brown DC (1971) Close-range camera calibration. Photogrammetric engineering, pp 855–866

Cho W, Schenk T, Madani M (1992) Resampling digital imagery to epipolar geometry. In: International archives of photogrammetry and remote sensing, Washington, D.C., USA, vol XXIX.B3

Clarke TA, Fryer JG (1998) The development of camera calibration methods and models. Photogramm Rec 16(91):51–66

Clarke TA, Wang X, Fryer JG (1998) The principal point and CCD cameras. Photogramm Rec 16(92):293–312

Fraser C (2013) Automatic camera calibration in close range photogrammetry. Photogramm Eng Remote Sens 79(4):381–388

Grussenmeyer P, Khalil OA (2002) Solutions for exterior orientation in photogrammetry: a review. Photogramm Rec 17(100):615–634. https://doi.org/10.1111/j.1477-9730.2002.tb01907.x

Hemmleb M, Wiedemann A (1997) Digital rectification and generation of orthoimages in architectural photogrammetry. In: International archives of photogrammetry and remote sensing, Göteborg, Sweden, vol XXXII.5C1B, pp 261–267

ISPRS (2018) ISPRS - the international society for photogrammetry and remote sensing. http://www.isprs.org/. Accessed 05 Feb 2018

Lowe D (1999) Object recognition from local scale-invariant features. In: Proceedings of the international conference on computer vision. https://doi.org/10.1109/ICCV.1999.790410

Luhmann T (2018) PhoX - photogrammetric calculation system. https://iapg.jade-hs.de/phox/. Accessed 22 Apr 2018

Luhmann T et al (2013) Close-range photogrammetry and 3D imaging, 2nd edn. De Gruyter, Berlin, 684 pp. ISBN: 978-3-11-030269-1

MiddleburyCollege (2015) Middlebury stereo datasets. http://vision.middlebury.edu/stereo/data/. Accessed 25 Apr 2018

Murtiyoso A, Grussenmeyer P, Börlin N (2017) Reprocessing close range terrestrial and UAV photogrammetric projects with the DBAT toolbox for independent verification and quality control. ISPRS Arch Photogramm Remote Sens Spat Inf Sci XLII-2/W8:171–177

Murtiyoso A et al (2018) Open source and independent methods for bundle adjustment assessment in close-range UAV photogrammetry. Drones 2(3)

OpenCV (2016) Camera calibration with OpenCV. https://docs.opencv.org/2.4/doc/tutorials/calib3d/camera_calibration/camera_calibration.html. Accessed 23 Dec 2019

Punmia BC, Jain AK, Jain AK (2005) Surveying III - higher surveying, 15th edn. Laxmi Publications (P) Ltd., New Delhi, 275 pp

Remondino F, Fraser C (2006) Digital camera calibration methods: considerations and comparisons. In: International archives of photogrammetry and remote sensing, Dresden, Germany, vol XXXVI.5

Trinder J, Fritz L (2008) Historical development of ISPRS. In: Zhilin L, Chen J, Baltsavias E (eds) Advances in photogrammetry, remote sensing, and spatial information sciences: 2008 ISPRS congress book. ISPRS book series. CRC Press, London, pp 3–20. ISBN: 978-0-415-47805-2

Tsioukas V (2007) Simple tools for architectural photogrammetry. In: International archives of photogrammetry and remote sensing, Athens, Greece, vol XXXVI-5.C53

Tsioukas V, Stylianidis E, Patias P (2000) Epipolar images for close range applications. In: International archives of photogrammetry and remote sensing, Amsterdam, The Netherlands, vol XXXIII.B5

Ullman S (1979) The interpretation of structure from motion. Proc R Soc Lond Ser B Biol Sci 203:405–426

Chapter 7
Production: Generating Photogrammetric Outcomes

Abstract The last chapter discussed issues related to the generation of photogrammetric products. For example, DEM/DSM/DTM and orthoimages are two key products of digital photogrammetry. Digital imaging is the origin for generating photogrammetric outcomes. The basic radiometric and geometric properties of digital imaging are discussed. Image matching techniques towards 3D object reconstruction is a key target in photogrammetry. The various matching techniques are deliberated while the role of interest points and operators are analyzed. In addition to that, dense image matching algorithms are also considered in this chapter, both from open source and commercial point of view. Point cloud and its processing is also discussed. Orthoimage production workflow is analyzed, as a result to deliver. In fact, it is an image geometrically corrected, in terms of camera tilt, lens distortion and object relief.

7.1 Introduction

Regardless their origin (terrestrial, aerial, satellite), images have been one of the dominant sources for obtaining geospatial information. Inevitably, the booming of sensor technology and the transition from analogue photographs to digital images spotlight the need for automating all processes connected to the generations of products from imagery. During the last decades, research and development focused on developing procedures, algorithms and tools on automation issues, such as orientation of images, DEM/DTM/DSM, building or road extraction, and many more. One of the most important issues towards this automation was the automatic finding of homologue points in overlapping images, through image matching techniques. In fact, this is one of the topics of this chapter as dense image matching is a hot topic. *3D* reconstruction using dense image matching enables the automatic extraction of *3D* models.

DEM/DSM/DTM and orthoimages are two key products of digital photogrammetry or what is known as photogrammetric production. They are important layers of information and are used in many applications. In aerial photogrammetry, there are many applications in mapping, urban planning, utilities, transportation, etc.

E. Stylianidis, *Photogrammetric Survey for the Recording and Documentation of Historic Buildings*, Springer Tracts in Civil Engineering,
https://doi.org/10.1007/978-3-030-47310-5_7

In close-range photogrammetry, there are also several applications in archaeology, architecture, civil engineering, etc. Besides, these products should be up to date, and thus a fast and financially viable production workflow is required. In addition to that, photogrammetric production must be inline with a quality control pipeline so as to guarantee the accuracy and quality of the outcomes, which means that the products are expected to be free of systematic errors and outliers.

Especially for historic buildings, which is the main interest of this book, DSM or *3D* point clouds are necessary for the *3D* reconstruction and modelling of the building. This is a vital step towards the recording and documentation process. In addition to that, orthoimages of building façades, specific architectural features, clearly defined parts of the building of exceptional importance, inside or outside the building, and many more, are integral part of the orthoimage photogrammetric production, towards the preservation of a building.

7.2 Digital Image

In the physical world, any quantity able to be measured through time over space or any higher dimension can be considered as a signal. As a result, a signal is a mathematical function; it conveys a piece of information, and it can be one, two or higher dimensional signal. A digital image is a two-dimensional signal which is defined by the mathematical function $f(x, y)$ where x and y are the two coordinates horizontally and vertically. Each point of the image defined by these coordinates is call *"picture element"* or *pixel*, and the value of $f(x, y)$ at any of these points, i.e. the pixels, gives the pixel value at that point of the image.

In practice, a digital image can be obtained in two fundamental ways. First, is the direct way, i.e. by using a digital camera or a sensor, where its size and shape determine the pixel size as well. The second way is the indirect one, where an analogue photograph from the old-fashion of films is scanned properly to produce the digital image. One way or another, a digital image is presented as a matrix I consisting of $i = 1, 2, ..., m$ columns and $j = 1, 2, ..., n$ rows. Each one of the matrix elements, i.e. a pixel, carries a value, an intensity value or a pixel value. Depending on the type of the image, the matrix consists only of one layer (a grey tone image) or several layers (colored image), as illustrated in Fig. 7.1.

While the digital image is sampled and mapped as a grid of pixels, each pixel is assigned a value (black, white, gray, color). These values are represented by zeros and ones in binary code. The binary digits, which are called "*bits*", for each pixel are accumulated in a sequence by a computer which is used to depict the image. The bits are then interpreted and read by the computer for further production.

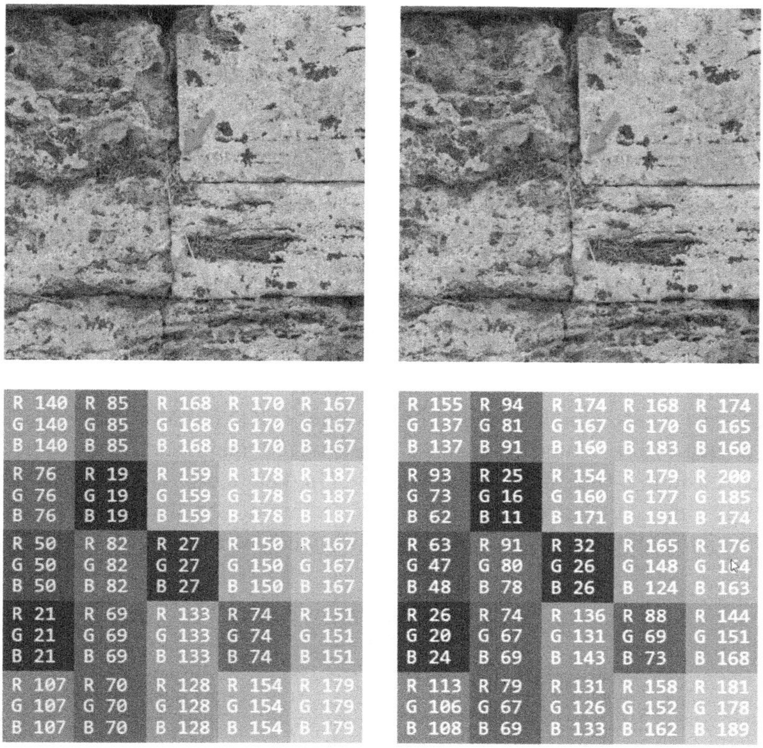

Fig. 7.1 Examples of grayscale and colored image and the matrices of pixels values

7.2.1 Radiometric Properties of Digital Image

Bit depth is determined by the number of bits used to describe every pixel. The greater the bit depth, the greater the number of tones that can be depicted, either grayscale or color. Digital images can be generated in black and white, grayscale, or color mode. A black and white, i.e. a bitonal image, is represented by two tones, black (0) and white (1) pixels, and thus every pixel consists of 1 *bit* each. A grayscale image is represented by 2 to 8 bits or more, and a color image is represented by a bit depth ranging from 8 to 24 or higher. In a 24-bit image, the bits are often divided in (3 per channel): 8 for red, 8 for green, and 8 for blue, and the combinations of those bits are generating other colors. The following binary calculations for the number of tones represented by common bit depths are helpful to understand this term:

1 bit = $2^1 = 2$ tones.
2 bit = $2^2 = 4$ tones.
3 bit = $2^3 = 8$ tones.
4 bit = $2^4 = 16$ tones.

8 bit = $2^8 = 256$ tones.
16 bit = $2^{16} = 65{,}536$ tones.
24 bit = $2^{24} = 16.7$ million tones.

If 256 levels are used, where 0 is the black and 255 (0) is the white, then the image is described with a radiometric resolution of $8 - bit$ and thus the intensity value is represented by an $8 - bit$ number. Photogrammetric automatic measurements in colored images can be performed by converting the gray scale images, and thus all calculations are taking place in the intensity level.

With respect to the radiometric properties of an image, there are two terms and statistical quantities which are very important and useful for either the image pre-processing before applying the automatic photogrammetric calculations or for post-processing towards the final outcome's generation.

First is *brightness* which reflects to mean value of all pixel values across the image. It is a degree of pixel intensity; if most pixels of the image carry small values -for instance, close to 0- the image is called dark. On the contrary, if most image pixels have large values, e.g. close to the maximum value of 255 for an 8-bit image, the image is called bright. Practically, brightness is calculated by the following formula (7.1):

$$f_m = \frac{1}{r \cdot c} \sum_{x=0}^{r-1} \sum_{y=0}^{c-1} f(x, y) \tag{7.1}$$

where

f_m is the mean gray value.
r, c is the number of rows and columns respectively in the image.
$f(x, y)$ is the gray value of the pixel at position (r, c).

The second term is *contrast* which reflects to the standard deviation of all pixel values in the image. In this case, it is the degree of scattering between the pixel values, and it can be determined as the difference between the brightest and the darkest pixels in the image. If the image has low contrast, it appears discolored, and the details in the image are barely observable. An image with high contrast is sharp, and the details can be easily recognized. The formula for image contrast calculation is:

$$\sigma = \frac{1}{r \cdot c} \sum_{x=0}^{r-1} \sum_{y=0}^{c-1} (f(x, y) - f_m)^2 \tag{7.2}$$

where

σ is the standard deviation, i.e the contrast of the image.
r, c is the number of rows and columns respectively in the image.
$f(x, y)$ is the gray value of the pixel at position (r, c).
f_m is the mean gray value as calculated in Eq. 7.1.

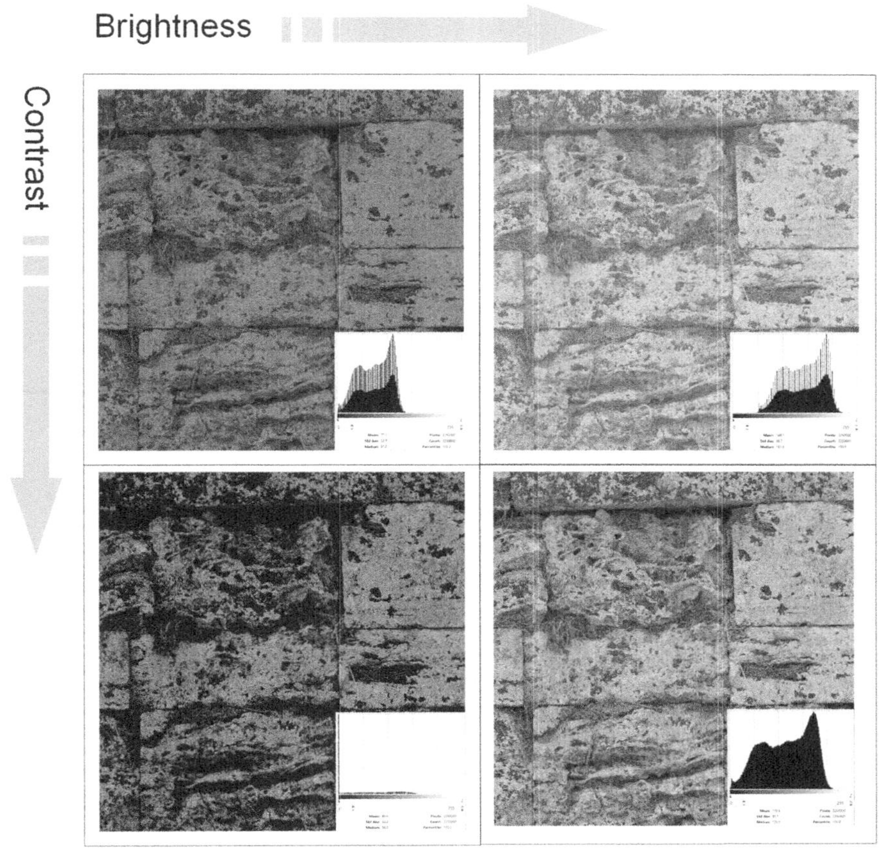

Fig. 7.2 Brightness and contrast in a snapshot on an image presenting a wall

As an image is a matrix with distinct values, the frequency of each image gray value can be represented by an *image histogram*. Frequency is rarely equal for all gray values of the image. Figure 7.2 illustrates the behavior of an image histogram while adjusting its brightness and contrast. Histogram equalization generates a uniform histogram and, in that way, compensates the contrast in the image. Depending the case, before, during or after the photogrammetric production, the processing of an image histogram is extremely helpful.

Image entropy is a quantity that describes the distinctive nature of the uncertainty of a gray value in a digital image. It is equal to the number of bits needed for saving a gray value of one pixel, or in image compression with the amount of information which must be coded by a compression algorithm. Image entropy is measured in bits per pixel value, and it is calculated by the following formula (7.3):

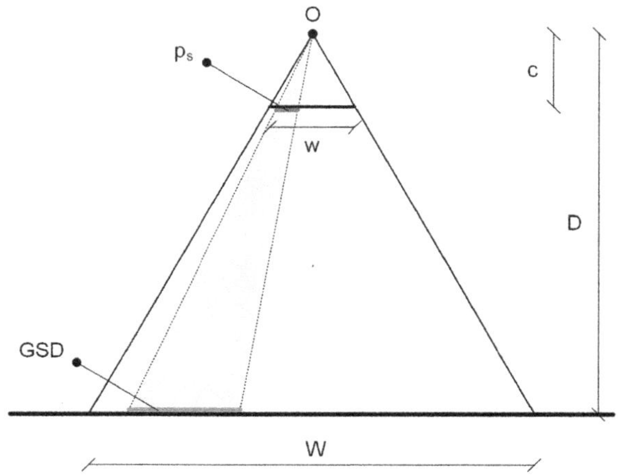

Fig. 7.3 GSD calculation in photogrammetric image acquisition

$$Entropy = -\sum_{i=0}^{f_{max}} p_i \cdot \log_2 p_i \tag{7.3}$$

where

f_{max} is the maximum gray value in the image.
p_i is the probability of a gray value f_i occurrence in the image.

Images with low entropy, for instance the "black images" (e.g. the upper left image in Fig. 7.2), contain a lot of black pixels, and thus they have small contrast in size. A total gray image has a zero entropy. High entropy images have a very good balance of contrast from one pixel to the next (e.g. the image at the lower right in Fig. 7.2). Low entropy images can be compressed much more than the high entropy images.

Entropy as a term originated from thermodynamics as (Shannon 1948) applied it to the field of information theory. In fact, the consideration is for the information content of a particular pixel of a variable as $-p \cdot \log_2 p$, where p is the probability of the variable taking on the value. As the $\log_2 p$ is negative, there is a minus (-) sign in Eq. 7.3.

7.2.2 Geometric Properties of Digital Image

The *pixel size* or geometric resolution is one of the basic properties of a digital image. It is an important feature that affects the accuracy of the image measurements and therefore the photogrammetric outcomes. An object must cover 2–4 pixels in order to be recognized during image measurements. Even though the pixel is located in the image plane, it is important to have a better sense of its real meaning in the object.

This is possible by projecting the pixel in the object space. In this case, the *ground sampling distance (GSD)* is equal to the size of the pixel projected on the object space as illustrated in Fig. 7.3.

The calculation of GSD is based on the following formulas, as extracted from the features in Fig. 7.3:

$$\frac{c}{D} = \frac{w}{W} = \frac{p_s}{GSD}$$
$$GSD = \frac{D}{c} \cdot p_s \tag{7.4}$$

In addition to GSD, the width of the area covered in the object space can be calculated as well, as following:

$$W = \frac{D}{c} \cdot w \tag{7.5}$$

where

c is the principal distance.
D is the distance between the image and the object.
w is the width of the camera sensor.
W is the width of the area covered in the object space.
p_s is the pixel size.
GSD is the ground sampling distance.

Another important parameter concerning geometric properties of digital images is base b between 2 consecutive images. It is the necessary distance between 2 serial images in order to cover an area in the object space with a specific overlap (Fig. 7.4). This is very useful in close-range photogrammetry and terrestrial images taken each

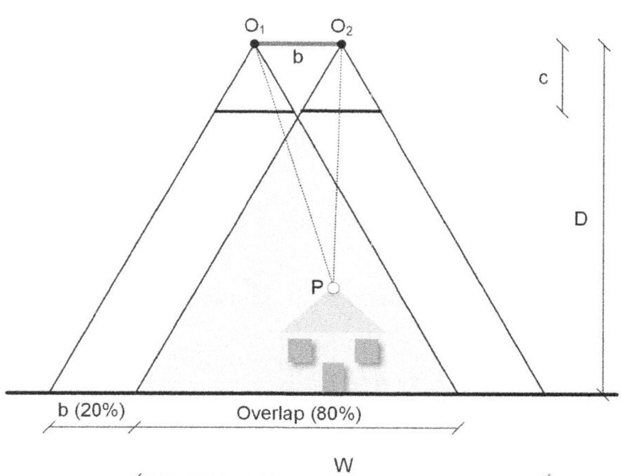

Fig. 7.4 The base (distance) between two consecutive images

time. In this case, distance b between 2 consecutive images is calculated by the following formula (7.6):

$$b = (1 - \frac{Overlap\%}{100}) \cdot W \tag{7.6}$$

Image *resolution* is the ability to recognize fine spatial details. The usual and proper indicator of resolution is the so-called dots-per-inch (dpi) which expresses resolution for digital images. In fact, it is a number of points (or pixels) that are rendered in 1 $inch$. This means that the greater the number of dots, the greater the density and thus the higher the resolution of the image. This also reflects to the pixel size. For instance, if the resolution of an image is 600 dpi, this means that 600 dots can fit in 1 $inch$, i.e. in 2.54 cm. This way, the pixel size of this image can be calculated as following (7.7):

$$\frac{0.0254m}{600} = 0.0000423m = 42.3\mu m \tag{7.7}$$

The dimensions of an image are the horizontal and vertical measurements of the image features expressed in pixels. The image dimensions in pixels can be calculated by multiplying the height and the width of the image by the proper dpi. In addition to that, a digital camera also has pixel dimensions, expressed as the number of pixels in the horizontal and vertical dimensions that define its resolution, for example a 24.3 megapixels camera (6000 $\times$ 4000). A photograph of 5 $\times$ 7 $inches$ scanned at the resolution of 600 dpi, the dimensions of the image in pixels is:

$$\begin{aligned} 600 \times 5 &= 3000 \\ 600 \times 7 &= 4200 \\ Dimensions &= 3000 \times 4200 \end{aligned} \tag{7.8}$$

Another digital image property is the *image file size*. It is calculated by the multiplication of the surface area of ($height \times width$) to be scanned with the bit depth and the dpi^2. As image file size is given in bytes, constituted of 8 bits, the previous figure is divided by 8, and thus the general formula is (7.9):

$$Filesize = \frac{height \times width \times bitdepth \times dpi^2}{8} \tag{7.9}$$

If the pixel dimensions are known, then the formula is (7.10):

$$Filesize = \frac{pixelheight \times pixelwidth \times bitdepth}{8} \tag{7.10}$$

As the images are directed for the necessary processing in photogrammetry, and file size is an important aspect in the calculations, the following naming conventions should be always considered, especially when handling huge images in the photogrammetric workflows:

1 Kilobyte (KB) = 1,024 bytes.
1 Megabyte (MB) = 1,024 KB.
1 Gigabyte (GB) = 1,024 MB.
1 Terabyte (TB) = 1,024 GB.

7.3 Image Matching Techniques

As *3D* object reconstruction is a key target in photogrammetry, measuring homologous points in two or more images is one of the most common and challenging tasks in photogrammetry. Especially in close-range photogrammetry where the arbitrary and unconventional type of image configuration is almost the usual case. In photogrammetry, automatic discovery of the homologous points in other images is known as *image matching* while in computer vision as *correspondence problem*.

By finding automatically the position of a point in two or more images, this directly implies that in we captured the image coordinates of this point in as many images as it appears. Hence, it is an important step towards the determination of its *3D* coordinates through the collinearity equations. It is easy to perceive the intended meaning that this condition directly affects and accelerates the computational processes in photogrammetry.

In image matching, all is about finding a correspondence in a pattern. A matching entity, which can be considered as a template, is compared with a patch in other images. A similarity measure is introduced to illustrate the quantitative measure of the matching and evaluate the match status of the entities. Until recently, the matching of all image pixels was considered a time-consuming process due to the limitations of the processing power. Nowadays, it is feasible, and it will be analyzed next (Sect. 7.4).

However, performing a "pixel by pixel" matching in the images has a tremendous calculation cost, considering the massive mathematical determination over the images. Moreover, these calculations may lead to uncertainties due to occlusions or repetitive patterns, i.e. repetitive existence of pixels with same or identical gray values. For these reasons, image matching is classified as an ill-posed problem, which means that it does not have a unique solution to the correspondence problem.

An ill-posed problem may not have a solution or not have a unique solution.

- For a point in the left image, the homologous point of the right image may possibly not detected, for instance, because of a hidden point or shadows. This is known as an occlusion problem.
- It is possible to have a maximum correlation/matching values in more than one points due to structures (e.g. bricks). This is known as a repeated pattern problem.
- The solution may be unstable due to the presence of noise in the image resulting from the poor texture. This is known as a poor texture problem.

In order to transform image matching from an ill-posed to a well-posed problem, additional information and constraints, like the examples given below, can be added to assist the matching process succeed.

- The lighting is stable during image acquisition.
- Use of the same spectrum in the 2 or more images used for matching.
- The object under study is stable during image acquisition.
- The surface of the object is opaque.
- It is helpful if the approximate average relief of the object is known.
- The approximate coverage between the images is known.
- The parallax is changing smoothly, which reflects to a smooth relief.
- The structure of the object is planar.

The scope of this section is to provide an overview of the basic methods for automatic measurements in digital images, especially area-based matching, feature-based matching and dense image matching.

7.3.1 Area-Based Matching

In area-based matching (ABM), the primitives considered for the correspondence problem are the (gray) values in the image pixels. As a single pixel is almost impossible to strongly describe the exact position of a point, several neighboring pixels are examined so as to overcome this ambiguity matter. To do so, instead of using a single pixel from the one (left) image, an image *patch* is extracted from this image, to be searched in the second (right) image. This image patch in the first image is called a *template*. Typically, the template is squared, i.e. equal number of pixels in width and height, since the position of the template is traced in the central pixel. This template is moved across the other image, and it is compared with patches of the same size.

It is obvious that moving this template along the entire width and height of the image would dramatically increase the computational cost in finding the homologous point. Usually, a close to the actual, but not completely accurate, position of the corresponding point is derived from the orientation parameters of the stereo pair. For instance, this could happen if the relative orientation between the two images is known. In this case, the comparison is limited from the whole image to a smaller area which is called the *search area*.

As the (moving) template is transposed from pixel to pixel across the search area in the second image, a value of the chosen similarity measure is calculated. It is supposed to be the case that the corresponding or homologous point in the center of the template to be located in that position where the appropriate value of the similarity measure is determined. This is depending on the similarity measures of specific features. Cross-correlation and least squares matching are the most useable area-based matching techniques in photogrammetry.

A larger template demands more singularity of the matching entity. However, matching of a larger template can be affected by the orientation of the images or the geometric distortions generated by the object relief. The singularity criterion is difficult to be achieved in the cases of repeated patterns or low contrast, or even in homogeneous-textured areas, for instance in the case of a building colored monotonously plain. ABM is able to find conjugate points regardless if an area a point belongs is a hidden one, and thus occluded areas should also be excluded.

In any case, ABM is an intriguing issue, and as such, the size and the location of the search area are important parameters during the matching process in order to keep away from mismatches.

The values of orientation parameters, the object relief that are nearly but not exactly correct and hierarchical approach are used under normal conditions to tackle this issue. A hierarchical approach is a coarse-to-fine strategy where the image matching process uses an image pyramid. An image pyramid is a set of images where the geometric resolution is reduced by 2, from level to level, both in rows and columns (Figs. 7.5, 7.6). The easiest way to calculate the pixels at the next lower pyramid level is by eliminating every second row and column of the current resolution. Nevertheless, in order to avoid big losses of information when changing to a lower resolution, the pixel values are derived from the pixel values of the higher resolution by means of a Gaussian or binomial filter.

A coarse-to-fine strategy points to a matching process that is starting at higher levels of an image pyramid, where small features are suppressed. The parameters determined in a higher level of the image pyramid are then transposed to a lower level of the image pyramid and used as starting point for matching at this level. In

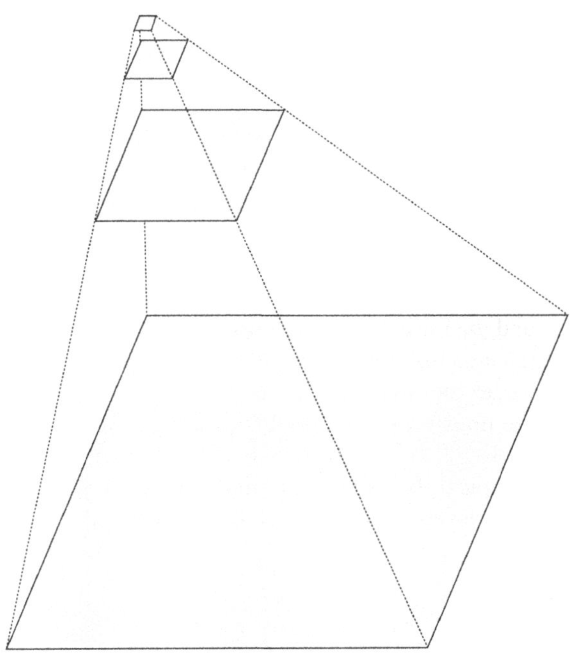

Fig. 7.5 The principle generating image pyramids

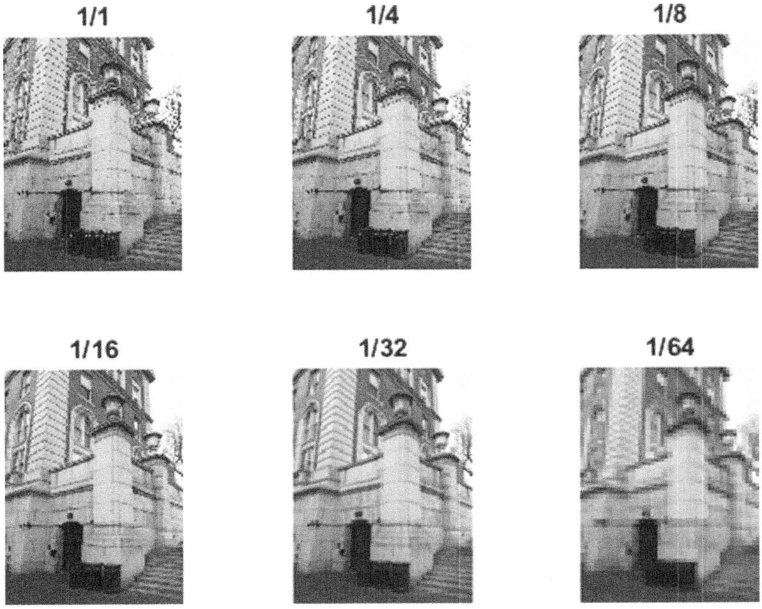

Fig. 7.6 An image pyramid example

the level where the geometric resolution is of high quality, the approximate values of the determined parameters are good to locate the search window. In this case, the image matching methods end up with a subpixel accuracy.

As introduced previously, when working with an image stereo pair, it is possible to use additional geometric constraints by considering epipolar geometry. These constraints can be applied along the epipolar lines and are helpful not only to reinforce matching capabilities but also to reduce computational cost.

Figure 6.33 shows how the concept of the epipolar line constraint can be used in image matching. The epipolar lines are formed at the intersections of the epipolar plane with the 2 image planes. The epipolar plane is determined by the 2 perspective centers O and O' and an object point P_1. For that reason, the homologous points p_1 and $p_1^{'}$ must lie on the corresponding epipolar lines el and el'. This way, the search area becomes much smaller while searching for the point. However, in order to render the image matching process along the epipolar lines easier, the images can be transformed to normalized images, where all epipolar lines in the image are parallel.

A typical example of an image pyramid is provided in Fig. 7.6. Several image pyramid levels from 1/64 to 1/1 are illustrated.

Correlation

In this case, *correlation coefficient r* is the measure of similarity between the template *f(x,y)* and the patch window *g(x,y)*. The place where the best result is located is supposed to be the position of the template window in the image. The calculation of the correlation coefficient is based on the covariance and standard deviations as following (7.11):

$$r = \frac{\sigma_{fg}}{\sigma_f \cdot \sigma_g} \tag{7.11}$$

The analytical expression of equation (7.11), for the correlation coefficient is:

$$r = \frac{\sum_{x=0}^{r-1}\sum_{y=0}^{c-1}(f(x,y)-\bar{f})\cdot(g(x,y)-\bar{g})}{\sqrt{\sum_{x=0}^{r-1}\sum_{y=0}^{c-1}(f(x,y)-\bar{f})^2\cdot\sum_{x=0}^{r-1}\sum_{y=0}^{c-1}(g(x,y)-\bar{g})^2}} \tag{7.12}$$

where

r is the correlation coefficient.
σ_{fg} is the covariance of the pixel values in the template and patch window.
σ_f, σ_g are the standard deviations of the pixel values in the template and patch window respectively.
$f(x, y), g(x, y)$ are the pixel values in the template and patch window respectively.
$\bar{f}, \bar{g}$ are the pixel mean values in the template and patch window respectively.
r, c are the number of rows and columns in the template and patch window (typically the same).

The correlation coefficient is bounded in the range of values $-1 \leq r \leq 1$. The values close to 1 indicate that the template and patch window are likely matched, while values close to 0 indicate no similarity. Values close to -1 point out that the positive and the negative of an image are matched.

The chosen template is transferred from pixel to pixel over the search window, and a correlation coefficient is determined in each position. The position where the correlation coefficient *r* is calculated with the highest value is potentially the position of the best match of the template in the search area.

The example in Fig. 7.7 illustrates the mechanism of image matching based on finding the maximum of the correlation coefficient r. As a template, a cross was chosen covering an area of 39×39 $pixels$, and as a search area, a little bit bigger, i.e. 51×51 $pixels$. While the template is transferred from pixel to pixel over the search area, the correlation coefficient is determined in each position as shown in the middle of this figure. The graph on the right shows values of the representation of the recorded correlation coefficients. The position where the correlation coefficient *r* is calculated with the highest value, i.e. 0.91 at $row = 25$ and $column = 26$, is the position of the best match of the template in the search area.

The correlation coefficient does not directly provide information about the accuracy of the best matched found position. However, the reliability of a determined

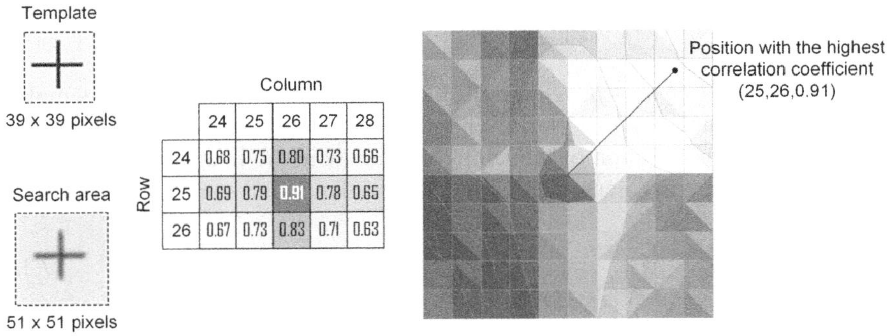

Fig. 7.7 Example of image matching with correlation coefficient determination

position depends on the radiometric properties of both the template and the search area. It is obvious that they can vary due to the different illumination and angle view, temporal changes, or projection of the images. As already discussed, the standard deviation (Sect. 7.2.1) of the pixel values and the entropy are measures of contrast for the images and can be used for the evaluation of chosen template for matching.

If subpixel accuracy is necessary in cross correlation, additional calculations should come after. In this case, the values of the correlation coefficient around its maximum are approached by a continuous function, e.g. a polynomial of second order, like the one given in 7.13:

$$r_a = r_c + v = a_0 + a_1 \cdot r + a_2 \cdot c + a_3 \cdot r \cdot c + a_4 \cdot r^2 + a_5 \cdot c^2 \tag{7.13}$$

where

r_a is the adjusted correlation coefficient.
r_c is the calculated correlation coefficient.
$a_1, a_2, \ldots, a_5$ are the coefficients of the 2nd order polynomial.
r, c are the pixel coordinates.
v are the residuals.

In order to find the maximum of this correlation function, the first derivative of this function should be set to zero. The coefficients of Eq. 7.13 are determined by least squares adjustment. Usually, a window of 3×3 or 5×5 seems to be optimal for fitting the second order polynomial.

Least-Squares Matching

Correlation coefficient is one method towards finding homologous points between two images. It gives good results, but it is not the perfect choice for measuring the similarity between the template and the patch window in the search area, due to their geometric and radiometric differences. The requirement to perform the correlation matching to subpixel accuracy drive a number of authors: Förstner (1982), Ackermann (1984), Grün (1985), Rosenholm (1987), Zheltov and Sibiryakov (1997), considering the so-called least squares matching (LSM).

Since then, LSM found tenths of applications both in terrestrial and aerial terrestrial photogrammetry, while the LSM algorithm has been used in a large number of photogrammetric software packages. The main idea behind LSM is the minimization of differences in pixel values between the template and the patch window in the search area, following a least squares adjustment process where geometric and radiometric corrections are considered.

The term "*adaptive*" (Grün 1985) was assigned to LSM because it allows the automatic changing of the number of parameters and weighting observations, depending on their state of being significant and the solution numerical stability. In order to have a successful LSM, an approximate position of the potential homologous point within the search area is necessary. It should be in the area of few pixels, and for this reason correlation can be used, as already explained.

Suppose two homologous image regions, i.e. the one expressed by the template window $f(x,y)$, and the other expressed by the patch window $g(x,y)$ in the search area. Ideally, if there were no errors, these two functions should have been equal. Due to the noise in both images, an error function $e(x,y)$ is introduced as shown in 7.14:

$$f(x, y) - e(x, y) = g(x, y) \tag{7.14}$$

The noise function is caused by the different radiometric and geometric effects in both images. The goal is to find the position of $g(x,y)$ in the search area, and by this way to determine the match point. This goal is succeeded by finding the optimum geometric and radiometric transformation parameters of one of the windows that is minimizing the error. The distances between the pixel values in the template and patch window should be minimized. LSM is a non-linear adjustment problem, and thus due to the geometric and radiometric transformation, Eq. 7.14 must be linearized.

As (Grün 1996) analytically explains, Eq. 7.14, after considering both the geometric and radiometric transformation, becomes as next:

$$\begin{aligned} f(x, y) - e(x, y) = \{ & g^o(x, y) + g_x \cdot da_{11} + g_x \cdot x_o \cdot da_{12} + g_x \cdot y_o \cdot da_{21} + \\ & g_y \cdot db_{11} + g_y \cdot x_o \cdot db_{12} + g_y \cdot y_o \cdot da_{21} + \\ & r_s + g^o(x, y) \cdot r_t \} \end{aligned} \tag{7.15}$$

a_{ij}, b_{ij} are the affine transformation parameters.
r_s, r_t are the 2 radiometric parameters, shift and scale, respectively.

As the original Eq. (7.14) is non-linear, the final solution is obtained iteratively.

7.3.2 Featured-Based Matching

While ABM functions directly on the pixel values, feature-based matching (FBM) uses as a spot from which can develop the extracted features from the images, such as points, edges, or regions. These specific features are based on steep changes in the pixel values, which many times agree almost exactly to scenic boundaries in the object space as well.

According to (Förstner 1986), FBM consists of three steps:

1. Selection of distinct features, such as points, edges, etc., separately for each image.
2. Generation of a preliminary catalogue of candidate pairs of corresponding features, based on the selected similarity criterion.
3. Produce a final catalogue of paired features, consistent with an object model.

Interest Points and Operators

Features of interest in an image, such as corners, edges, etc., are of high importance in photogrammetry, as they are highly connected with the automation of processes such as point cloud and DSM generation. The need for accurate and consistently good in quality automatic surface extraction remains a key milestone towards image-based modelling. In order to apply image matching techniques and produce accurate results, the identification of well-described interest points is a requirement.

An *interest point* is that point at which the direction of the object boundary changes abruptly. It may also be an intersection point between two or more edge segments (e.g. a window corner).

Interest operators are basically algorithms that detect such features of interest in an image. Various interest point operators have already been proposed in the international literature. Each operator has different characteristic parameters; however, they depend on pixel (gray) values within the searching window.

Moravec (1977, 1979) was the first who developed the idea of using "*points of interest*", i.e. points with special characteristics that can be detected easily. These points are described as points that become noticeable when high intensity differences occur in every direction. Actually, it is based on the assumption that the interest points have high variances in all directions. In his approach, Moravec calculates through this operator the mean square sums of gradients in the 4 principal directions of a window. If this value is greater than a threshold set, there is a characteristic feature in the image that has notable variations in the intensity values in the 4 selected directions.

The general structure of the Moravec operator calculations is given in the following Eq. 7.16.

Fig. 7.8 Interest point detection with Moravec interest point operator

$$MO = min \begin{Bmatrix} \sum[g(r, c) - g(r, c+1)]^2 \\ \sum[g(r, c) - g(r+1, c)]^2 \\ \sum[g(r, c) - g(r+1, c+1)]^2 \\ \sum[g(r, c+1) - g(r+1, c)]^2 \end{Bmatrix} \quad (7.16)$$

Figure 7.8 shows the results from applying Moravec operator in a building façade image.

Harris and Stephens (1988) proposed what has become well-known as the *Harris* or as *Plessey* feature point detector. It was developed 11 years after the Moravec operator, trying to overcome the limitations of the Moravec operator and generate a more concrete feature point detector. A different form of this operator was discussed and tested in a relevant study (Stylianidis 2003).

The windowed neighborhood autocorrelation for each point *E(x,y)*, in an input image *I*, where a local window *w(x,y)* is moving over the image, is defined by the following formula (7.17):

$$E(x, y) = \sum_{i=-n}^{n} \sum_{j=-m}^{m} w(x, y) \times |I(x+i, y+j) - I(x, y)|^2 \quad (7.17)$$

This function detects the local maxima in all directions of the autocorrelation function. However, this is very sensitive to noise in the image, particularly near the edges.

In their approach, (Harris and Stephens 1988) proposed the use of a second-order development of *I(x,y)* and a Gaussian weighting function *G* to decrease its sensitivity to noise and suppress anisotropy. This way, function *E* becomes (7.18):

$$E(x, y) = \sum_{i=-n}^{n} \sum_{j=-m}^{m} \frac{1}{2 \cdot \pi \cdot \sigma^2} \cdot e^{-\frac{(i^2+j^2)}{2 \cdot \sigma^2}} \times \left| x \cdot \frac{\partial I}{\partial x} + y \cdot \frac{\partial I}{\partial y} \right|^2$$
$$E(x, y) = (x, y) \cdot A \cdot (x, y)^t \tag{7.18}$$

where

$$A = G(\sigma) * \begin{pmatrix} \left(\frac{\partial I}{\partial x}\right)^2 & \frac{\partial I}{\partial x} \cdot \frac{\partial I}{\partial y} \\ \frac{\partial I}{\partial x} \cdot \frac{\partial I}{\partial y} & \left(\frac{\partial I}{\partial y}\right)^2 \end{pmatrix} \tag{7.19}$$

The positive-definite matrix *A* has eigenvectors v_{max} and v_{min} associated with the eigenvalues λ_{max} and λ_{min} convey information about the directions of the local intensity changes. The vector v_{max} is linked with rapid changes in intensity and is perpendicular to the edge. The vector v_{min} is linked with low changes in intensity and is parallel to the edge. When the two eigenvalues are small, the pixel belongs to a homogeneous zone, when one eigenvalue is high, it is an edge, and when both eigenvalues are large, it is a corner.

In order to overcome costly calculation of eigenvalues, the inventors of this operator proposed the use of the following measure:

$$\begin{aligned} E(x, y) &= det(A) - k \cdot (trace(A))^2 \\ E(x, y) &= \lambda_{min} \cdot \lambda_{max} - k(\lambda_{min} + \lambda_{max})^2 \end{aligned} \tag{7.20}$$

The function *E(x,y)* is positive and high for a corner. The parameter *k* is defined empirically, and it is set at 0.04. The interest points detected by this operator correspond to the local maxima of the function *E(x,y)*. This corner detector is invariant to translations, rotations and affine intensity variations; however, it is affected by scale and affine transformations. An example from the implemented algorithm is given in Fig. 7.9.

One year earlier, another interest point operator was introduced. The well-known *Förstner* operator (Förstner and Gülch 1987) as it was established to be said. It is extensively adopted in photogrammetry and used in the computer vision community for many years, and at the present time it is still used in many 3D object reconstruction applications. Förstner operator was developed with the aim of detecting and precisely locating points of interest, such as corners and centers of circular image features, and the ultimate goal to be used in photogrammetric image matching applications.

Fig. 7.9 Interest point detection (100 strongest) with Harris/Plessey interest point operator

An example from the implementation of Förstner operator in a building façade image is given in Fig. 7.10.

However, (Förstner 1982) laid the foundation for developing an operator for interest point detection, suitable for robust automatic image matching. Generally, such points are considered the points of circular shape that have intensity differences from the adjacent ones. The least square estimation provides a variance-covariance matrix where the position uncertainty could be extracted; i.e. the eigenvalues λ_1, λ_2 indicate the semi-axes of an error ellipse. This error ellipse is directly related to the image content, and in linear features it is elongated along the line, while it is a small circular ellipse in well-defined points.

Förstner (1986) first introduced an operator for identifying points suitable for automatic alignment, points that would be strong in their coincidence. Generally, such points are considered to be circular in shape and have a marked intensity difference from their adjacent ones. The minimal-square coincidence returns a variance-covariance table resulting in the lack of position uncertainty. This deficiency is directly related to the content of the image, and in linear phenomena, it is elongated in the direction of the line, while in point phenomena, it approaches a circle.

According to the normal equations of least squares estimation, the variance-covariance matrix between 2 matching windows is given by:

$$cov\begin{pmatrix} d\hat{x} \\ d\hat{y} \end{pmatrix} = \sigma^2 \cdot \begin{bmatrix} \sum g_x^2 & \sum g_x \cdot g_y \\ \sum g_x \cdot g_y & \sum g_y^2 \end{bmatrix}^{-1} = \sigma^2 \cdot Q = \sigma^2 \cdot N^{-1} \qquad (7.21)$$

Fig. 7.10 Interest point detection with Förstner operator

where g_x and g_y are the intensity differences in x and y directions. However, using Eq. 7.21 it is possible to calculate the error ellipse parameters (7.22).

$$q = \frac{4 \cdot det(N)}{tr^2(N)} = \frac{4 \cdot \lambda_1 \cdot \lambda_2}{(\lambda_1 + \lambda_2)^2} = 1 - \left(\frac{\lambda_1 - \lambda_2}{\lambda_1 + \lambda_2}\right)^2 \tag{7.22}$$

where

λ_1, λ_2 are the eigenvalues of matrix N.
$det(N)$ is the determinant of matrix N.
$tr(N)$ is the trace of matrix N.

The shape of a good point should be as circular as possible, i.e. q is closer to 1. The measure of the ellipse size w is given by Eq. 7.23:

$$w = \frac{det(N)}{tr(N)} = \frac{1}{tr(N)} \tag{7.23}$$

The experiments show that a "*Förstner point*" is observed if thresholds in w and q are exceeded. The threshold for w is $[0.5 \ldots 1.5] \times w_{mean}$, where w_{mean} becomes as the mean of w for the overall image. The threshold for q is $[0.5 \ldots 0.75]$, while the suitable windows to used 5×5 to 7×7 pixels.

Smith and Brady (1997) introduced the *SUSAN* operator. A patent has been granted for the SUSAN operator. It was developed as a new approach to low-level image processing and more specifically, in edge and corner detection and structure preserving noise reduction. SUSAN algorithm is an accurate, noise-resistant and fast edge and corner detector that uses non-linear filtering to describe exactly which segments of the image are intimately related to each single pixel. Each pixel has associated with it a local image region which, resembling without being identical, brightness to that pixel.

According to the inventors of the operator, SUSAN is looking for areas of similar brightness, and as a result for points of interest within a weighted circular window. They introduced an algorithm that signifies the search window central pixel as the nucleus. The area inside the window that has similar intensity values to the nucleus is calculated, and it is called as the univalue segment assimilating nucleus (USAN). When USAN has a low value, this reflects to a corner, as the central pixel would be very dissimilar from the pixels around it. The local minima of the smallest USAN continue to exist as valid interest points, after evaluating the results and getting rid of the outliers, while the comparison between pixel brightness values is figured out, deploying the following Eq. 7.24:

$$c(\mathbf{r}, \mathbf{r}_o) = \begin{cases} 1 \rightarrow if\ |I(\mathbf{r}) - I(\mathbf{r}_o)| \leq threshold \\ 0 \rightarrow if\ |I(\mathbf{r}) - I(\mathbf{r}_o)| > threshold \end{cases} \tag{7.24}$$

where

$\mathbf{r}$ is the position of any other point inside the circular window.
$\mathbf{r}_o$ is the position of the nucleus in the image.
$I(\mathbf{r}$ is the brightness value of any pixel.

The comparison is performed with full awareness for each pixel inside the circular window. The total number of pixels possessing similar brightness values as the nucleus is found by the sum of Eq. 7.25:

$$n(\mathbf{r}_o) = \sum c(\mathbf{r}, \mathbf{r}_o) \tag{7.25}$$

The calculated value of $n(\mathbf{r}_o)$ is measured against a geometric threshold, which is set as g. The SUSAN algorithm applies threshold values in order to recognize which features are suitable and which are not, i.e. to detect features that make acceptable interest points and non-acceptable features. For detecting a corner in the image, the threshold value g is set to half of the maximum value of $[n(\mathbf{r}_o), n_{max}]$, and if n_{max} is less than threshold g, then a corner has objective reality.

SIFT was introduced by Lowe (2004) as a key point detector to extract features from images taken from different viewpoints and under different illumination conditions. In fact, SIFT stands for "scale invariant feature transform", which means that even if there are images of different sizes, of different viewpoints and of different depth, i.e. the images have different scale, SIFT is able to overcome this issue, as it is invariant to the scale of the images. It is patented by the University of British

Columbia, which means if anybody wants to use SIFT for commercial purposes, they have to pay.

By using SIFT as a key point detector in an image-based modelling workflow, the goal is to:

- Extract distinctive invariant features, which then can be correctly matched against a large number of features from many images.
- Have invariance to image scale and rotation.
- Be robust in affine distortion, change in *3D* viewpoint, in addition of noise, and change in illumination.

The advantages from the application of SIFT key point detector can be summarized as following:

Locality: The features are local, which means that the algorithm is robust to occlusion and clutter.
Distinctiveness: Individual features can be matched to a large number of objects.
Quantity: Many features can be identified, even for small objects.
Efficiency: Performance close to real-time.

SIFT algorithm can extract stable features with sub-pixel accuracy and match them, considering 128-dimension descriptors.

According to the SIFT inventor, in order to detect stable features in the images with different sizes, the algorithm generates image pyramids and utilizes a staged filtering approach based on the Difference of Gaussian (DoG) function, instead of Laplacian of Gaussian (LoG). The reason is that DoG is much faster. Image pyramid space consists of two components. The first is Gaussian scale space, in which each octave is calculated by convolution between a down-sampled image with different Gaussian kernels, such as (7.26):

$$L(x, y, \sigma) = G(x, y, \sigma) * I(x, y) \tag{7.26}$$

where

$G(x, y, \sigma)$ is a variable-scale Gaussian.
$*I(x, y)$ is an input image.

The convolution method with Gaussian kernels can limit the noise effect and meliorate the features' invariance when the scales of the images are altered.

The second part is DoG scale space, i.e:

$$D(x, y, \sigma) = (G(x, y, k \cdot \sigma) - G(x, y, \sigma)) * I(x, y) = L(x, y, k \cdot \sigma) - L(x, y, \sigma) \tag{7.27}$$

where k is a constant multiplicative factor.

DoG, if it is compared with LoG, has lower computational costs and roughly the same function to extract constant features.

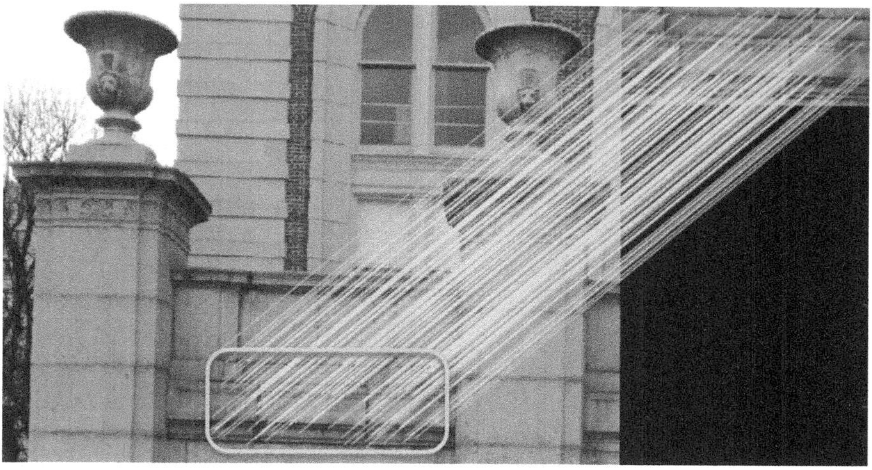

Fig. 7.11 Interest point detection with SIFT operator

It has been widely used in photogrammetry and engineering geodesy applications. Lingua et al. (2009) executed a performance analysis of SIFT in aerial and close-range photogrammetric applications and also introduced and validated an auto-adaptive version of SIFT operator (A^2SIFT). Reiterer et al. (2010) presented an image-based measurement system in engineering geodesy which uses key point detectors. Sun et al. (2014) presented an extended version of SIFT (L^2SIFT) to extract and match features from large images, like in the case of large-scale aerial photogrammetry.

Figure 7.11 provides an example on how SIFT is working and what is the output. The matching points between the image A and image B, i.e. a small part of the first image which is highlighted in the rectangle, are connected with straight lines.

FAST operator was developed by Rosten and Drummond (Rosten and Drummond 2006) as a high-speed feature detector, appropriate for real-time frame-rate applications. The inventors of FAST showed that the algorithm is able to process live Phase Alternating Line (PAL) video at a full frame rate using less than 7% of the available processing time, in comparison to the Harris operator (120%) and SIFT operator (300%). In addition to that, the operator performed better than the other existing operators with respect to the ability of detecting the same corresponding interest points from different viewpoints.

It functions similarly to the SUSAN algorithm. FAST operator considers a small image patch and judges if it "looks" like a corner or not. Like in SUSAN, a circular window is scanned across the image, and the pixels' intensity values inside or around the window are compared to that of the central pixel. According to the FAST investors, the aim of the operator was to improve SUSAN, in terms of speed and invariance to rotation, and transformation and changes in scale. In addition to that, inductive machine-learning methods that extract patterns from large datasets are utilized in the FAST corner detection. FAST operator is quicker and more credible, and it is chosen in many applications.

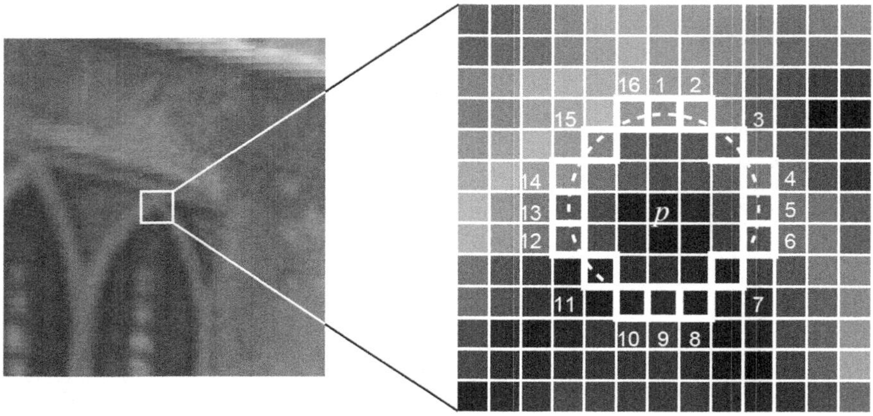

Fig. 7.12 The 16-pixel circular window used by FAST operator. (Source: Rosten and Drummond 2006)

According to Rosten and Drummond (2006), the algorithm functions by considering a circular window of 16 pixels around the corner candidate point *p* (Fig. 7.12). The point is considered of interest when a set of *n* contiguous pixels in the circular window are all brighter than the intensity I_p of the candidate pixel *p* plus a threshold *t*, or all darker than $I_p \leq t$. For each one of the 16 locations on the circular window $x \in 1 \ldots 16$, the pixel at that location relative to *p*, denoted by $p \rightarrow x$, can have one of the 3 states:

$$S_{p\rightarrow x} = \begin{cases} d, & I_{p\rightarrow x} \leq I_p - t \quad (darker) \\ s, & I_p - t < I_{p\rightarrow x} < I_p + t \quad (similar) \\ b, & I_p + t \leq I_{p\rightarrow x} \quad (brighter) \end{cases} \tag{7.28}$$

For each *x* in the circular window, $S_{p\rightarrow x}$ is computed for all points $p \in P$, where *P* represents the set of all pixels in all training images and divides *P* into 3 subsets, namely P_d, P_s or P_b. Each candidate interest point *p* is assigned a relevant value $PS_{p\rightarrow x}$.

Consider a boolean variable K_p, which is true if *p* is an interest point, otherwise is false. The algorithm selects a *x* value which yields the most information about whether the candidate pixel is an interest point, where the *x* value is chosen by measuring the K_p entropy. This starts with computation of the entropy *H(P)* (7.29) of *K* for the set *P*.

$$H(P) = (c + \bar{c}) \cdot \log_2(c + \bar{c}) - c \cdot \log_2 c - \bar{c} \cdot \log_2 \bar{c} \tag{7.29}$$

where

$$\begin{aligned} c &= \left|\{p \mid K_p \ is \ true\}\right| (number \ of \ corners) \\ \bar{c} &= \left|\{p \mid K_p \ is \ false\}\right| (number \ of \ non - corners) \end{aligned} \tag{7.30}$$

Fig. 7.13 Interest point detection (100 strongest) with FAST operator

As soon as the x value providing the most information has been chosen, the procedure is put in practice recursively on all the three P subsets. The procedure stops when the entropy of a subset is zero; i.e. when all p values in the subset have the same value as K_p, they are either are all interest points or non-interest points. This procedure ends up with a decision tree which categorizes all the detected points. After that, it is transformed to C-code and compiled twice for optimization and used as a detector of corner points.

FAST algorithm computes a score function V (7.31) for each detected interest point. The score function is the sum of the absolute difference between the pixels in the contiguous arc and the center pixel.

$$V = max\left(\sum_{x \in S_{bright}} \left|I_{p\to x} \le I_p\right| \le t, \sum_{x \in S_{dark}} \left|I_p \le I_{p\to x}\right| \le t\right) \tag{7.31}$$

As a result, a non-maximal suppression is applied to remove points that have an adjacent point with a higher V.

The score value can be used to take out interest points below a selected threshold. The points with the highest absolute difference between the pixels in the contiguous arc and the center pixel p could be considered as high-quality interest points.

A typical example from the output of running FAST operator in a building façade image is illustrated in Fig. 7.13.

Fig. 7.14 Interest point detection (100 strongest) with SURF detector

SURF (speed-up robust features) is a scale- and rotation-invariant interest point detector and descriptor introduced by Bay et al. (2006, 2008). According to the inventors of SURF, the most valued feature of an interest point detector is its repeatability, that is to say, whether it reliably finds out the same interest points under different viewing conditions. The detector is non-commercial.

It uses the maxima of the Hessian matrix determinant to detect interest points. Hessian matrix is a way to package all the information of the second derivatives of a function. The calculation of the Hessian matrix is accelerated by approximating the underlying Gaussian filter process with simple box filters of 9×9.

SURF descriptor characterizes the distribution of the intensity content within the neighborhood of the interest point by building on the distribution of first order Haar wavelet responses in x and y direction rather than the gradient, and uses only 64 dimensions. As SURF is a scale-invariant interest point detector, the input image is analyzed at different scales in order to secure invariance to scale changes.

Figure 7.14 illustrates an example of the interest points detected using SURF detector.

Indicative examples from studies comparing and evaluating the various feature point detectors are from Rodehorst and Koschan (2006), who are evaluating the performance of 3 feature point detectors, namely SUSAN, Plessey and Förstner, and Jazayeri and Fraser (2010) who are testing the Förstner, the SUSAN and the FAST operator, in terms of which is the optimal for feature-based matching in convergent close-range photogrammetry.

7.3.3 Image Content Enhancement

An image is a snapshot at a very specific time, and recording and documentation of cultural heritage, based on image-based modelling techniques, considers images as the main input. As a result, image content plays a catalytic role in many photogrammetric processing and feature extraction methods. The content and its quality will guide image-based processing and feature extraction algorithms to provide stable and accurate outcomes. There are several enhancement algorithms to process the image properly and increase its quality (Gonzalez and Woods 2002; Jähne 2005; Pratt 2014).

Image enhancement algorithms are able to increase the visual appearance of the images; either it is necessary for the photogrammetric processing pipeline or for the final delivery of the outcomes towards recording and documentation cultural heritage. They can further improve the quality of details, which is quite important especially in the case of feature extraction. But the final documentation products should be also of high fidelity.

In image-based *3D* reconstruction workflows, low-texture surface is a real problem to be faced. It is met frequently in historic buildings with the plaster or clay building facades or any other material which creates low-texture surface. This type of problems causes difficulties to feature detection methods, but also to image matching algorithms, leading to erroneous matching outcomes, and thus to deficiency in delivering high fidelity *3D* representations of the historic buildings.

Wallis (1976) filter is amongst the tools used within the photogrammetric community in order to reinforce the image content and facilitate photogrammetric image processing (Baltsavias 1991; Jazayeri and Fraser 2010; Gaiani et al. 2016). In practice, it is a digital image processing filter that further improves the quality of the contrast levels and makes flat the different exposure to attain the desired objective, i.e. the similar brightness of pixel gray values. It is targeting local dynamic range correction and edge-based denoising, which is very important in the photogrammetric workflows.

According to Pratt (2014), Wallis (1976) suggested the following form for the Wallis operator:

$$
\begin{aligned}
G(i,j) = \; & [F(j,k) - M(j,k)] \cdot \frac{A_{max} \cdot D_d}{A_{max} \cdot D(j,k) + D_d} \\
& + [p \cdot M_d + (1-p) \cdot M(j,k)]
\end{aligned}
\tag{7.32}
$$

where

$F(j,k)$ is the input image.
$M(j,k)$ is the estimated mean value of the original image at point (j,k).
A_{max} is a maximum gain factor that prevents excessively large output values when $D(j,k)$ is small.
D_d is the desired standard deviation.
M_d is the desired average mean.

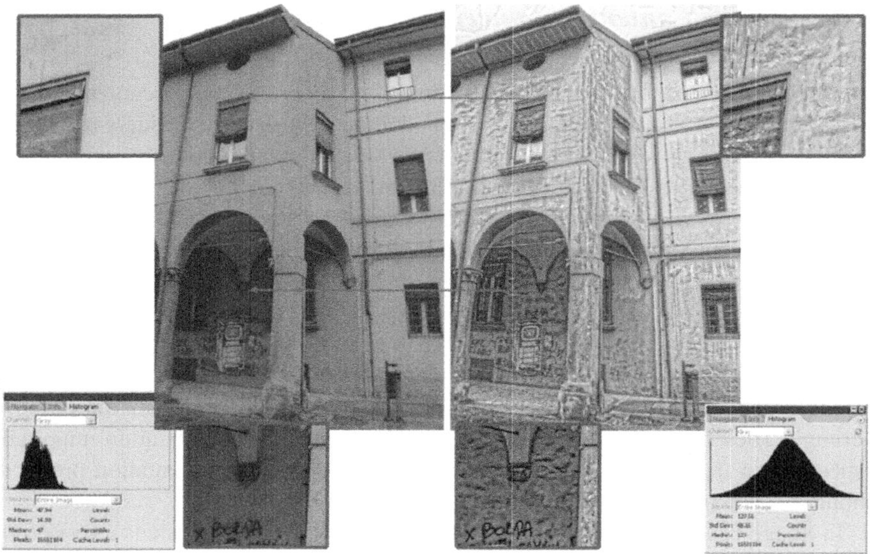

Fig. 7.15 Applying Wallis filter in an image (Gaiani et al. 2016)

$D(j, k)$ is the estimated standard deviation.

p is a mean proportionality factor controlling the background flatness of the enhanced image ($0.0 \leq p \leq 1.0$).

However, Gaiani et al. (2016) presented their own approach to implement the Wallis filter and generate the enhanced output image:

$$(x, y) = S \times [input(x, y) - m]/(s + A) + M * B + m \times (1 - B) \tag{7.33}$$

Based on their approach, they are suggesting the following steps for the implementation of the Wallis filter:

1. Let *S* be the standard deviation and *M* the mean of the input image.
2. For each (*x*,*y*) pixel in the image.
3. Calculate local mean *m* and standard deviation *s* using a $N \times N$ neighborhood.
4. Calculate the enhanced output (Wallis-filtered) image.

The quality of the enhanced output image, i.e. the Wallis-filtered image, depends on the values of two parameters (7.33). On the one hand, it is the *contrast expansion factor A*, while on the other, it is the *brightness forcing factor B*. The correct choice of the values for the parameters *A* and *B* is the main difficulty in applying Wallis filter. It is more an "ad-hoc" set of instructions than an easily applicable and automatic photogrammetric workflow. A Wallis-filtered image example is illustrated in Fig. 7.15, and it shows the results of Wallis filtering; lower image contents are boosted, whereas a better histogram is achieved (Gaiani et al. 2016).

7.4 Dense Image Matching

Image matching has a long history in photogrammetry, dated back to the use of analogue workflows and the 1950s, where the earliest matching algorithms were developed (Hobrough 1959). Since then, it is part of photogrammetric, computer vision and image processing applications; the most important of which is *3D* reconstruction and modelling. As the automation in image-based *3D* modelling digital workflows becomes more and more demanding, image matching algorithms are becoming more and more creative and inventive.

The various image matching techniques differ according to the number and the capacity of image-based information used. When all the pixels in the image are used in the matching process, then this is called *dense image matching* (DIM). DIM is a trend in image-based *3D* modelling entire workflow, as it aims at calculating a depth/height value for every single pixel of an image. This is extremely important because this makes easier the generation of accurate and highly regarded detailed DSMs. DIM of multiple overlapping images can generate surface models at a density and accuracy which could not be anticipated not long ago. In practice, DIM is able to perform a pixel-based matching in order to calculate the *3D* coordinates of a dense point cloud, i.e. to generate the so-called dense *3D* point cloud.

DIM is an alternative approach to obtain *3D* coordinates for an enormous number of points. In practice, through DIM, a corresponding point for nearly every pixel in the image is searchable. Rather than searching the whole image for feature points, DIM will compare two overlapping images row by row. Basically, this makes smaller down to one-dimension search, and the problem is much simpler. This is in need of an image rectification step previous to image matching. The images should be warped in such a way that each row of pixels in one image concur precisely to one row in the other image. In fact, this means that the rows of the images should be parallel to the epipolar line.

In this case, the algorithm can function row by row and pixel by pixel. For each pixel, the algorithm will look in the corresponding row for the pixel that is most probable to represent the same point in the object space. From the practical point of view, the algorithm will perform that by comparing the grey or color of the pixel and its neighbors. Simultaneously, a constraint is defined to make certain a specific amount of normality in the result. When a pixel is discovered in the second image that is a good match to the same pixel in the first image, the pixel position is recorded. As soon as two corresponding pixels are recognized, typical photogrammetric workflows can be employed to calculate the *3D* intersection for the pixel.

7.4.1 Semi-global Matching

According to an image-based approach, the delivery of a dense *3D* point cloud reflecting the configuration of the various stereo models, suppose that a high density image matching should be performed for each image pixel. Usually, additional

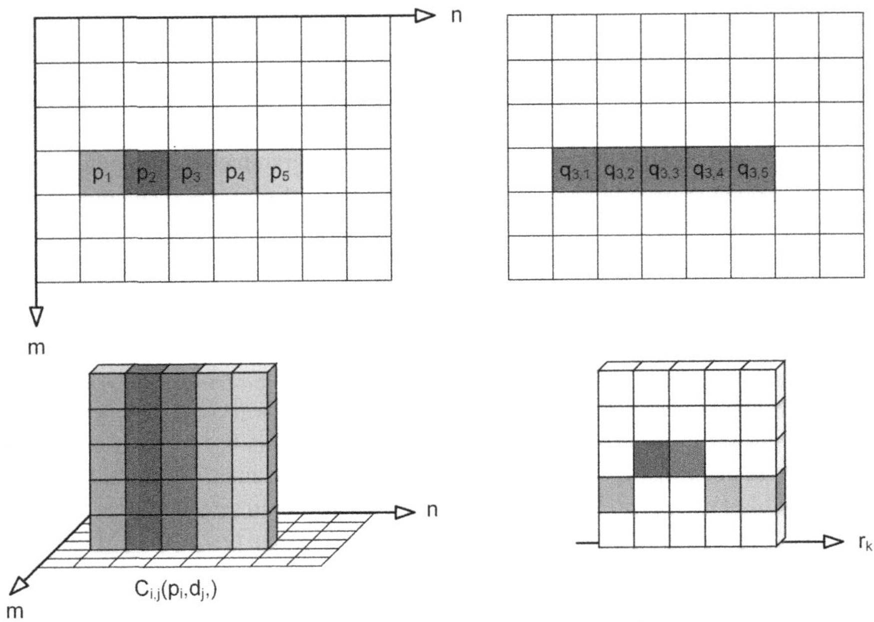

Fig. 7.16 SGM concept according to (Haala 2011)

constraints are introduced to encounter the overall uncertainties from such a huge processing effort for each pixel. According to (Szeliski 2011), stereo matching methods could be local or global. The local approaches use the pixel intensity values in a bounded region, calculate the disparity at a given point, and finally a smoothing and a local "winner-take-all" (WTA) optimization at each pixel. The global methods make smoothness assumptions and then use the energy-minimization framework to handle a global optimization problem.

Hirschmüller (2005, 2008) put forward the idea of semi-global matching (SGM) for making available a match for each image pixel, originating from computer vision domain. In the photogrammetric domain, (Haala 2011) argues that SGM is able to reduce significantly the computational complexities from the global approach. In his study, he approximates a global approach by minimizing matching costs, which are grouped into a class along a certain number of *1D* path directions through the image (Fig. 7.16). The particularity of this SGM approach makes the runtime rational, providing a dense *3D* point cloud, even using a dataset of large imagery.

This pixel-based matching approach is targeting to connect each pixel coordinate in one image to its corresponding pixel coordinates in the second image. By detecting the corresponding pixels between 2 or more images, the viewing rays of the matched object point can be intersected in the (object) space. Inevitably, this can lead to determine the *3D* coordinates of the object points. As the target is the recapturing of a surface information in a *3D* high density form, the matches are preferably calculated for every single pixel of the image.

However, every matching relation should fulfil some criteria to be fixed, or as it is called that every such relation has a matching cost to be considered in order to identify the right pixel to prevail. The aggregation of all costs describes exactly the global matching cost. This global matching cost should be minimal for the most favorable pixel matching. Haala (2011) suggests the minimization of an approximation of the global cost, which leads to a fast numerical solution. Different variants of the SGM algorithm have been presented from time to time in the photogrammetric community, for example, the multi-view DIM algorithm from Yan et al. (2016). In their study, they introduce such an algorithm which is based on a graph network. The proposed method is compared to other state-of-art methods, like SURE and PhotoScan.

7.4.2 Dense Image Matching Algorithms

The stunning developments in the information technology sector and the evolution of new image matching approaches unavoidably led to the development of software tools and solutions in the image-based *3D* reconstruction domain. Amongst the most widely triggered algorithms and software packages for the photogrammetric community, the following are considered (in alphabetical order): MicMac, PhotoScan, Pix4D, PMVS and SURE. They are originating both from commercial and open-source domain.

Different evaluations of these algorithms have taken place over time. Haala (2013) presents the results from a European Spatial Data Research Organisation (EuroSDR) benchmark on image-based DSM generation, which were processed by different groups with different software systems. Remondino et al. (2013) present a critical review and analyses of the DIM algorithms, both in open-source and commercial domain, while at the same time, they evaluate the performance and potential of these algorithms by using different datasets. The potential of DIM approaches for *3D* data captured from oblique airborne imagery was discussed by Cavegn et al. (2014). They evaluated the potential of DIM algorithms in a test bed provided by the ISPRS/EuroSDR initiative, while the test scenario is demonstrated using matching results from two software packages, PhotoScan and SURE. The comparison of 5 different sensors and 4 different software packages is performed by Niederheiser et al. (2016). The formation of this comparative experiment is carried out in a single application, the modelling of a vegetated rock face. Svensk (2017) studied 4 software packages, open-source and commercial, in forest inventory application, to evaluate them and generate point clouds in order to estimate parameters of the trees. In their study, Alidoost and Arefi (2017) used UAS images to test the capabilities of four different software packages in the sense of high-density *3D* point cloud and DSM generation, over a historical site.

MicMac

MicMac is a free open-source photogrammetric suite that can be used for detailed and accurate image-based *3D* reconstructions (IGN-ENSG 2018). Even though in the beginning it was developed at the National Institute of Geographic and Forestry Information (IGN) and the National School of Geographic Sciences (ENSG) to handle high-resolution stereo satellite imagery for *3D* surface reconstruction (Pierrot-Deseilligny and Paparoditis 2006), nowadays, it is used in an extended application spectrum and covers all acquisition imagery configurations (terrestrial, drone, aerial). As a result, MicMac can cover from small objects to cities or even in urban or rural areas. According to Rupnik et al. (2017), MicMac has a distinct feature compared to its "competitors": the user can intervene from the command line by adapting the available parameters. A set of extensible markup language (XML) files are accessible, permitting an expert user to choose the optimum parameters according to the case. In addition to that, MicMac is available as a library to be exploited by skilled users for further developments. By offering these parameterization features, MicMac is far from an attractive graphical user interface (GUI).

The simplified architecture of MicMac is illustrated in Fig. 7.17. It is organized in a number of modules, which are accessible through a common command (mm3d). The photogrammetric workflow starts from the original images, passes through the estimation of the orientation parameters, to the multi-view stereo image matching (MVSM) and the delivery of the *3D* surface model. MicMac performs a bundle block adjustment (BBA) which is solved with the Levenberg-Marquardt method. DIM in MicMac is possible with a particular form of semi-global and global algorithms.

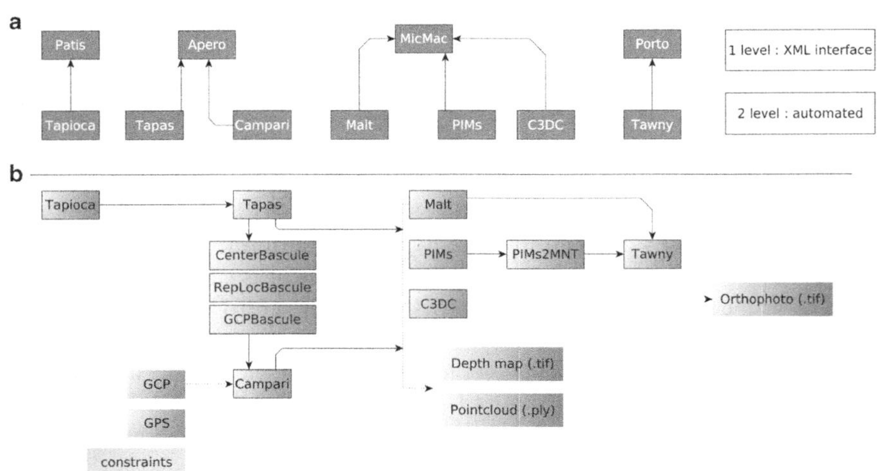

Fig. 7.17 MicMac architecture with low- & high-level modules (a) and processing workflow (b). (Source: (Rupnik et al. 2017))

Three different outcomes are delivered from MicMac: depth map, dense *3D* point cloud, and orthophoto (Rupnik et al. 2017).

Agisoft PhotoScan/Metashape

Agisoft PhotoScan, lately renamed to Metashape, is a stand-alone photogrammetric software solution for automatic processing of digital images into accurate spatial data, like dense point clouds, textured polygonal models, georeferenced true orthomosaics and DEMs. The software can be run on Windows, Linux and Mac OS X, both on a typical desktop configuration and on a multi-node cluster for massive data management.

The software is being developed by Agisoft (2019) LLC and is used worldwide as a profound yet easy-to-master solution for photogrammetric tasks in wide range of industries, from aerial surveying and mapping and precision agriculture to digital archaeology, to visual effects and game design.

The software workflow is linear and project-based, and it includes three main stages:

- photo alignment (auto calibration, AT, BBA).
- dense point cloud generation (dense matching).
- target output data generation (e.g. texturized *3D* model, orthomosaic, DEM, tiled model, etc.).

The user may interact with the software via intuitive GUI or run the software from command-line and automate the processing using Python scripts. Typically the project can be processed within a few hours, providing at the same time highly accurate results (up to 3 cm for aerial, and up to 1 mm for close-range imagery).

The software can use as input data not only still digital images, taken with consumer range and high dynamic range (HDR) cameras, but also video frames and multispectral images. It is capable of processing thousands of images, yet all the processing is performed locally. On the other hand, with floating license support, the software can be used on a virtual machine, thus allowing effective management of resources for hardware/software costs balance.

One of the key advantages of the software is its flexibility depending on the end user level of experience and complexity of the project. Even a non-professional can run the software following the instructions of step-by-step tutorials published on Agisoft web site (Agisoft 2019), while the advanced user is provided with a high-level of control over the processing. All settings can be altered manually and reviewed via software GUI as well as exported for further analysis in a detailed report.

Output data can be exported for further processing in external solutions or for publishing on-line and *3D* printing. The range of supported data formats includes the most popular OBJ and FBX formats for *3D* models, LAS for point clouds, GeoTIFF for DEMs and orthomosaics. LAS file import opens the way to combine LiDAR and photogrammetric data to benefit from both technologies' advantages.

Pix4D

Pix4D (2019) is an image-based software company delivering high-quality products by leveraging computer vision, photogrammetry, machine learning and artificial intelligence techniques. The company offers several products for real business problems.

The company was founded in 2011 and introduced Pix4D software. The software uses images or videos taken by hand, car or plane, and it is up to the user to create customizable results in a wide range of applications. In practice, it is a hybrid solution able to collect, process, analyze and share image-based data from mobile to desktop devices, and/or from cloud-based to on premise platforms.

Pix4D can process inputs originating from RGB, multispectral and thermal sensors while being camera agnostic.

The workflow of data processing with Pix4D consists of 3 steps that include all the basic photogrammetric procedures to calibrate images (internal and external orientations) and produce outputs such as densified point clouds, orthomosaics, *3D* textured meshes, DTMs and DSMs, etc.

The software is offered as a processing engine behind platforms that aim to scale up. The enterprise-ready application programming interface (API) products further open the field of possibilities. On demand, Pix4D provides Python-based software development kit (SDK) as well as API to their cloud solution in order to serve the needs of more complex services.

PMVS—CMVS—VisualSFM

PMVS is an MVS software that receives a package of digital images and camera parameters in order to perform and deliver the *3D* reconstruction of an object or a scene that is viewable in the images, only for rigid structure. It does not deliver a mesh model but only a set of oriented points. At each oriented point, the *3D* coordinate and the surface normal are computed (PMVS 2008).

CMVS uses the output of a SfM software as an input in order to disintegrate the input images into a set of image clusters of a reasonable and easy to handle size. In this case, MVS software can then be used to process each image cluster independently and in parallel. The *3D* reconstructions from all image clusters should not fail to capture any particular information that can be otherwise acquired from the entire image set. CMVS should be used in conjunction with an SfM software Bundler (2008) and an PMVS2 (PMVS version 2) (CMVS 2010).

VisualSFM is a GUI application for *3D* reconstruction using SfM. For dense *3D* reconstruction, this program combines the execution of PMVS/CMVS tool chain (VisualSFM 2018).

SURE

SURE (Photogrammetric surface reconstruction from imagery) is the Institute for Photogrammetry (ifp) at the University of Stuttgart, software for DIM (ifp 2016). Nowadays, SURE is distributed by the startup company nFrames (2018). It is a software solution based on the MVSM method, which delivers dense point clouds, i.e. one *3D* point per pixel is calculated, considering that a specified set of images and its orientations are available for use.

As a MVSM method, SURE uses single stereo models of 2 images which are processed and afterwards fused. The images become undistorted and rectified as stereo pairs. Then, the appropriate image pairs are carefully chosen and matched using a dense stereo method similar to the SGM algorithm. SURE does not use the original SGM method as introduced by Hirschmuller (Hirschmüller 2008); however, a more time and memory efficient solution achieving maximum productivity with minimum wasted effort is implemented. With respect to the original SGM approach, SURE searches the pixel correspondences using dynamic disparity search ranges and a tube-shape structure to store the costs of the potential correspondences (Fig. 7.16).

This stereo matching step is based on the library libTSGM, which executes a modified version of the SGM. The library consists of 3 main modules handling image rectification, DIM and multi-view structure computation (Rothermel et al. 2012). Stereo matching method leads to the disparity image which contains the results from the correspondence information. However, the target is to compute *3D* coordinates in the object space, and for this reason, a triangulation has to be performed. By using the orientation of the images, the rays are constructed corresponding to the pixel measurements and made to intersect properly in the object space. For this purpose, the viewing rays corresponding to the pixel measurements are constructed using the orientation in order to intersect them in space. In the successful matching scenario in a stereo pair, every pixel corresponds to one single and specific measurement, and thus 2 rays are intersected, guiding the algorithm to one single *3D* point for each matched pixel (Wenzel et al. 2013).

An example from the application of SURE DIM algorithm in an aerial imagery configuration from Warsaw city center in Poland is given in Fig. 7.18.

7.4.3 Point Cloud

The ultimate goal of photogrammetry is to succeed the determination of *3D* object point coordinates from photogrammetric image processing. Certainly, *3D* points should not be only accurately computed but also the suitable number of points so as to precisely describe the object. As already discussed, image matching procedures are leading to the correspondence of points in 2 or more images and thus finally to

Fig. 7.18 The results from SURE DIM algorithm. (Source: Kind courtesy of nFrames and Warsaw University of Technology)

the calculation of their *3D* coordinates. Especially, DIM deals with every single pixel of an image and the relevant determination of *3D* object point coordinates.

A set of data points in the *3D* object space is called *point cloud*. Although the term "point cloud" is apparently identified as an outcome that normally is produced by *3D* laser scanners, this is not the reality. A point cloud can be also produced by an image-based method, such as those photogrammetry makes a formal application. In practice, it is the basis for many other processes and outcomes, such as meshes, textured models, *3D CAD* models, rendering, visualization, animation, etc.

In principle, *3D* point clouds can be distinguished into two different types of point structures (Fig. 7.19):

- Sparse point cloud.
- Dense point cloud.

Many times, a point cloud should be aligned with other point clouds in order to create the overall model of the object. A simple example could be considered is

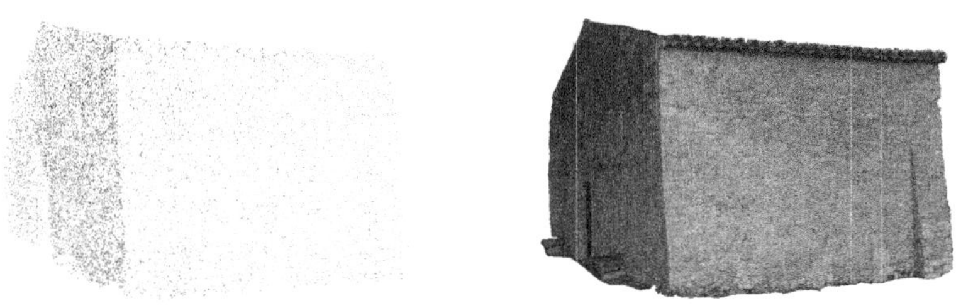

Fig. 7.19 Sparse (left) and the dense (right) 3D point cloud of a historic building

Fig. 7.20 A 3*D* model of a historic building

one point cloud for the outside part of a historic building and one point cloud for the inner part of the building. This is a prerequisite so as to generate a high fidelity *3D* model of the building. The process is called point cloud registration. However, common points are necessary between the 2 point clouds to be registered in a single and unified point cloud.

Usually, different platforms and sensors are employed to derive *3D* point clouds in historic buildings. *3D* reconstruction from UAV images are used to generate sparse or dense models mainly showing roof surfaces. In contrast to such sparse or dense models, 3D reconstructions from terrestrial images are much denser and show building façades and interior spaces. Moreover, LiDAR sensors could capture point clouds. *3D* point clouds do not only vary in the viewpoints, but also in their scales and point densities. As a result, the fusion of such *3D* point clouds with diverse origin is highly demanding.

Even though a dense point cloud looks like a "*3D* model", it is not considered as such. A *3D* model is the generated output of *3D* modelling, that is, the process of developing a mathematical representation of any surface of an object in *3D* space. Three-dimensional modelling is performed with the use of specialized software. Following a *3D* rendering process, a *3D* model can presented as a *2D* image (Fig. 7.20).

Meshing Point Clouds

Usually, *3D* point clouds do not keep their form, and they are converted to mesh models, polygon or triangle, through a series of actions which are very often alluded

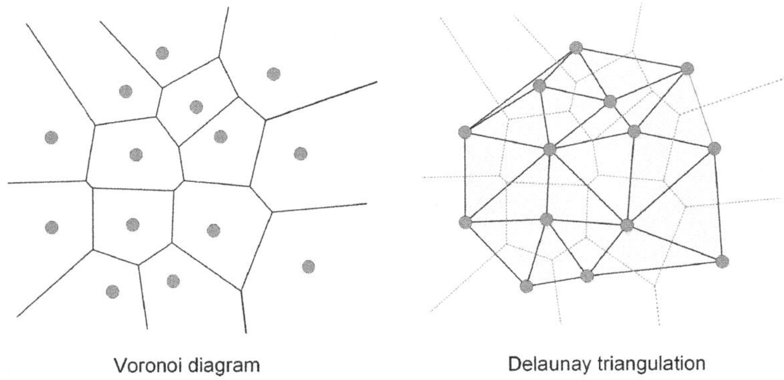

Fig. 7.21 The concept of Voronoi diagram and Delaunay triangulation

to *3D* surface reconstruction. The *3D* reconstructing of surfaces from point clouds is a topic maintained by careful and deliberated effort in computer graphics. There are many available techniques for transforming a *3D* point cloud to a *3D* surface. Two of the most widely used techniques are Delaunay triangulation and Poisson surface reconstruction.

In principle, as a process, *3D* reconstruction considering as input a *3D* point cloud, such as the outcomes of a dense image matching process, to a *3D* surface, is a fitting problem.

Delaunay triangulation, the dual shape of the Voronoi diagram, enables a unique division of the space points to be made based on the nearest neighbors (Fig. 7.21). In fact, it generates a network of triangles over the point cloud vertices.

For a given a number of points, the Voronoi diagram divides this space according to the nearest neighbors as it is shown in Fig. 7.21. The line between each couple of points is defined by the perpendicular bisector. According to Aurenhammer (1991), Voronoi diagrams are important because of 3 main reasons:

- They can arise in nature in various situations.
- They have interesting mathematical properties.
- They have proved to be powerful tools in solving seemingly unrelated computational problems.

Given a number of points, Delaunay triangulation is considered as that triangulation where none of the points is located inside the circumcircle of any generated triangle. Following this rule, it maximizes the minimum angle of all the triangles' angles in the triangulation.

Poisson surface reconstruction is another widely used *3D* surface reconstruction technique. In fact, it reconstructs a triangle mesh from a set of oriented *3D* points by solving a Poisson system, i.e. it is solving a *3D* Laplacian system with positional value constraints. Poisson systems are known widely for their resilience in the presence of incomplete data (Kazhdan et al. 2006; Kazhdan and Hoppe 2013). Poisson

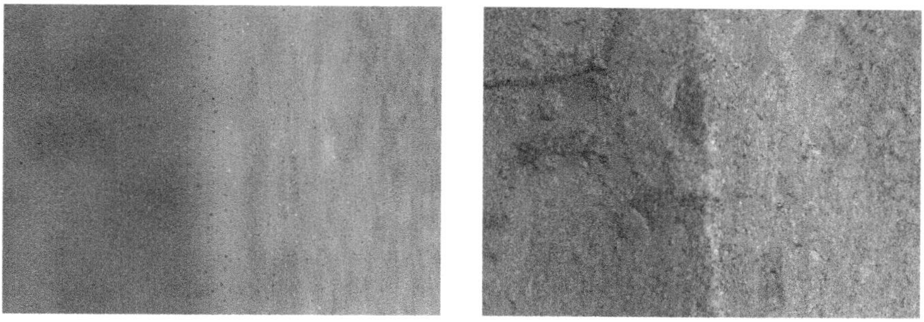

Fig. 7.22 Snapshots from the meshing results performed in the point cloud of a historical building from Fig. 7.19

reconstruction is considered very useful as a technique to provide a triangulated surface, especially when starting from noisy point clouds.

An example from meshing the point cloud of the historic building in Fig. 7.19 is given in Fig. 7.22. On the left is the result from the Poisson meshing after sampling the original point cloud, while on the right is the high fidelity mesh. Both images are snapshots from the corner of the building to provide a more detailed view of the outcomes.

FOSS for Mesh Handling

Point clouds are typically used to represent and measure an object's surface. They have many application areas, and the existing algorithms provide point cloud processing functionalities, such as downsampling, denoising, transforming point clouds, point cloud registration, geometrical shape fitting to point clouds, compare point clouds, etc. There are several free tools available, which someone can search and find freely available. However, 2 of the most widely used free tools within the photogrammetric, recording and documentation of cultural heritage community are MeshLab and CloudCompare.

MeshLab (ISTI-CNR 2016) is an open-source system, endowed with the prestigious Eurographics Software Award, for processing and editing 3D triangular meshes. It has been developed at the Visual Computing Lab of the Institute of Information Science and Technologies (ISTI), an institute of the Italian National Research Council (CNR). It makes available for use a set of tools for editing, cleaning, healing, inspecting, rendering, texturing and converting meshes. MeshLab renders features for raw data processing, produced by *3D* image-based or range-based tools and devices. It is also able to deliver models for *3D* printing (Cignoni et al. 2006).

MeshLab source code is also available, and it is distributed under the general public license (GPL) 3.0 licensing scheme, while also the homonym name of the system is a European Union intellectual property office (EUIPO) trademark owned by CNR. The logos are distributed under creative commons license (CC BY-SA 4.0), and they can be freely used inside any wikimedia project.

CloudCompare is a *3D* point cloud and triangular mesh processing free software. Originally, the software was designed and developed to compare dense *3D* point clouds, regardless of the origin of the raw data, i.e. image-based or range-based techniques. It was also developed to perform comparisons between a point cloud and a triangular mesh. Later, CloudCompare has been extended to a generic point cloud processing software including various algorithms, such as registration, resampling, statistics computation, interactive or automatic segmentation, etc. (CloudCompare 2018).

7.5 Orthoimage Production

An *orthoimage* is an orthorectified image. In fact, it is an image which has been geometrically corrected, in terms of camera tilt, lens distortion and object relief. In contrast to the single image rectification, in this case, the object is not assumed as a *2D* plane (reference), but as a real *3D* object with relief and a real irregular surface. An orthoimage is an image free from all sorts of errors and for this reason is a valuable source for real and correct measurements. Unlike the case of single image rectification, where only camera tilt and lens distortion are removed, in addition to that, an orthoimage is free of errors that are created from the relief, e.g. the topographic relief.

In summary, in an orthoimage, the following assumptions could be supposed to be the case:

- The formal perspective in the image has been eliminated, as a transformation from the central projection to orthographic projection is performed.
- The image displacements caused by the camera tilt and the object relief do not exist anymore.
- There is a uniform scale across the whole area covered by the orthoimage.
- Measurements can take place over the orthoimage, acting as a texturized and photographic map.
- Metric information can be extracted from the orthoimage by using a typical digitization procedure.

Figure 7.23 is illustrating graphically the transition from the central projection of an image to the orthographic projection of the orthoimage, as it is the normal case in regular maps. For an image, central projection generates perspective view. This is the way the images are generated through the entire camera system, including the lens. However, in the case of a map, and thus in orthoimages as well, the projection is orthographic, and consequently, the rays connecting the object space points with the points on the map/orthoimage are perpendicular to the projection plane. This also results in the orthographic representation of the objects being depicted perfectly vertical.

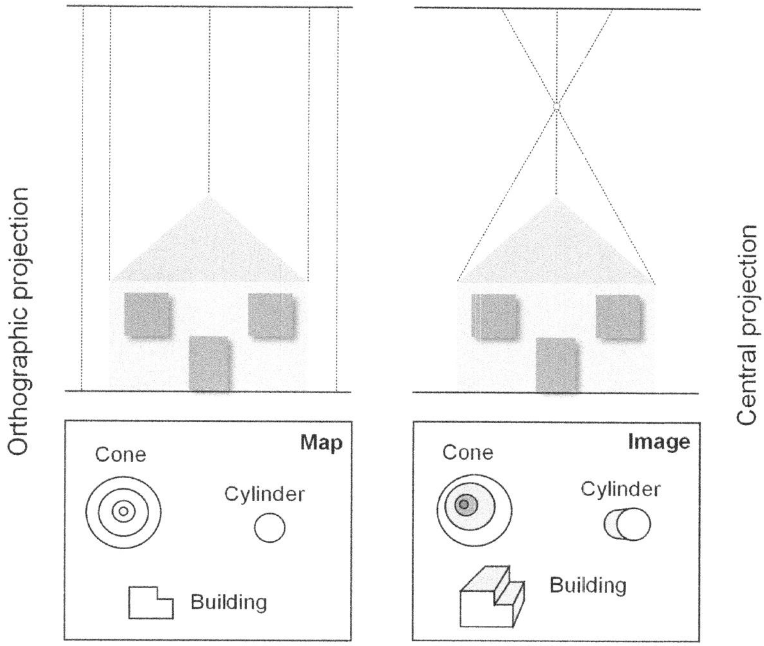

Fig. 7.23 The difference between central and orthographic projection; an image and a map

In order to produce an orthoimage, a workflow, like the one illustrated in Fig. 7.24, should be utilized.

In addition to that, the required data for orthoimage production are:

- The digital image. A grayscale or color image is a prerequisite. The pixel size and the GSD vary depending the camera used and the distance between the object and the camera. These parameters are affecting the overall final quality of the outcome.
- The interior orientation of the camera (including the camera calibration parameters).
- The exterior orientation of the image to be used in the orthoimage production process.
- The DSM which clearly describes the object *3D* model presented in the image.

Essentially, orthoimage production process transforms each pixel of the original image into an object space location. Usually, this is called a pixel georeference. The transformation can be applied in two ways: either from the original image to the transformed orthoimage, or vice versa, which actually is more convenient. As a result, and considering the graphical representation of the photogrammetric workflow for the production of an orthoimage in Fig. 7.24, the following explicit steps should be considered:

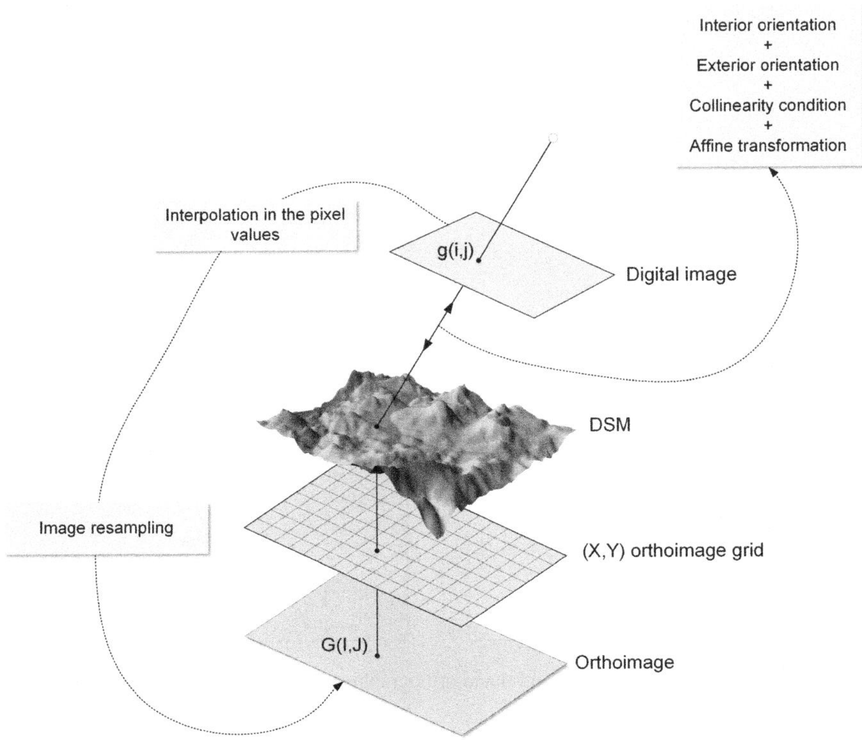

Fig. 7.24 Graphical representation of the orthoimage production workflow

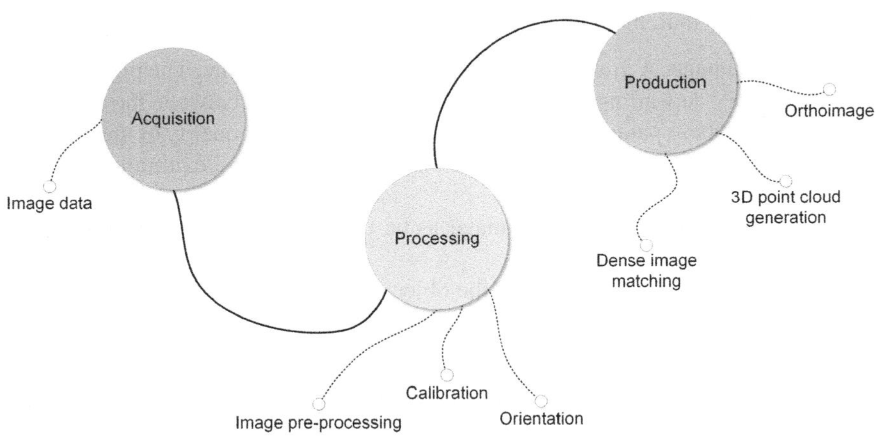

Fig. 7.25 The image-based method ecosystem

1. Define the area coverage of the orthoimage, e.g. upper left (X_{UL}, Y_{UL}) and lower right corner (X_{LR}, Y_{LR}), as well as the pixel size.
2. For each pixel (X, Y) which is now georeferenced and refers to the object coordinate system, perform an interpolation in the DSM to find the proper Z value for this specific pixel.
3. Implementation of the inverse collinearity equations to come up with the exact (x, y) coordinates of the pixel in the image plane.
4. Implementation of the inverse affine transformation to calculate the exact pixel coordinates (i, j).
5. Radiometric interpolation in order to compute the appropriate pixel value.
6. Attribute the calculated pixel value to the applicable pixel at the orthoimage.
7. Repeat the above steps for all pixels.

7.6 Summary—the Image-Based Method Ecosystem

Photogrammetry is a passive method as it is using passive sensors (cameras) to collect the primary information. It is an image-based method in contrast to (laser) scanning, which is a range-based method and belongs to active methods as it is using active sensors for the direct data acquisition.

In summary of the image-based methods, the following figure (Fig. 7.25) is illustrating the overall approach in the image-based ecosystem.

References

Ackermann F (1984) Digital image correlation: performance and potential application in photogrammetry. Photogramm Rec 11(64):429–439

Agisoft (2019). *Agisoft metashape.* Accessed: 2019-08-01. https://www.agisoft.com/

Alidoost, F. and H. Arefi (2017). "Comparison of uas-based photogrammetry software for 3d point cloudgeneration: a survey over a historical site".In *The International Archives of the Photogrammetry, RemoteSensing and Spatial Information Sciences*, Vol. IV-4/W4.https://doi.org/10.5194/isprs-annals-IV-4-W4-55-2017

Aurenhammer F (1991) Voronoi diagrams - a survey of a fundamental geometric data structure. ACM Computing surveys 23(3):345–405. https://doi.org/10.1145/116873.116880

Baltsavias EP (1991) Multiphoto geometrically constrained matching. In: PhD dissertation No 9561. ETH Zurich, Switzerland

Bay H et al (2008) Speed-up robust features (SURF). Comput Vis Image Underst 110(3):346–359. https://doi.org/10.1016/j.cviu.2007.09.014

Bay H, Tuytelaars T, Gool LV (2006) SURF: speed-up robust features. In Leonardis A, Bischof H, Pinz A (eds) Computer vision—ECCV2006. Graz, Austria, pp 404–416

Bundler (2008). *Bundler: structure from motion (sfm) for unordered image collections*". Accessed: 2018-05-22. http://www.cs.cornell.edu/~snavely/bundler

Cavegn S, et al (2014) Benchmarking high density image matching for oblique airborne imagery. In: The international archives of the photogrammetry, remote sensing and spatial information sciences, vol XL-3. https://doi.org/10.5194/isprsarchives-XL-3-45-2014

Cignoni, P. et al. (2006). "Meshlab: an open-source mesh processing tool". In: *6th Eurographics Italian chapter conference* Ed. by V. Scarano, R. D. Chiara, and U. Erra, pp. 129–136
CloudCompare, (2018) Cloudcompare. Accessed: 2018–05-21.http://www.cloudcompare.org
CMVS (2010). *Clustering views for multi-view stereo (CMVS).* Accessed: 2018-05-22. https://www.di.ens.fr/cmvs/
Förstner W (1982) On the geometric precision of digital correlation. In: International archives of photogrammetry and remote sensing, vol 24.3, Helsinki, Finland, pp 176–189
Förstner W (1986) A feature based correspondence algorithm for image matching. In: International archives of photogrammetry and remote sensing, vol 26.3/3. Rovaniemi, Finland
Förstner W, Gülch E (1987) A fast operator for detection and precise location of distinct points, corners and centres of circular features. In: Proceedings of ISPRS intercommission conference on fast processing of photogrammetric data, vol 437. Interlaken, Switzerland, pp 281–305
Gaiani M et al (2016) An advanced pre-processing pipeline to improve automated photogrammetric reconstructions of architectural scenes. In: Remote sensing 8.3. https://doi.org/10.3390/rs8030178
Gonzalez RC, Woods RE (2002) Digital image processing, 2nd Edn. Prentice Hall, Upper Saddle River, New Jersey 07458
Grün A (1996) Close range photogrammetry and machine vision, In: Atkinson KB (ed) Least squares matching: a fundamental measurement algorithm. Whittles Publishing, Dunbeath, pp 217–255
Grün A (1985) Adaptive least squares correlation: a powerful image matching technique. S Afr J Photogramm Remote Sens Cartogr 14(3):175–187
Haala N (2011) Multiray photogrammetry and dense image matching. In: Dieter F (ed) Photogrammetric week '11. Wichmann, Berlin/Offenbach, Germany, pp 185–195
Haala N (2013) The landscape of dense image matching algorithms. In: Dieter F (ed) Photogrammetric week '13. Wichmann, Berlin/Offenbach, Germany, pp 271–284
Harris C, Stephens M (1988) A combined corner and edge detector. In: Proceedings of the Fourth Alvey Vision Conference, vol 302. Manchester, UK, pp 147–151
Hirschmüller H (2005) Multiray photogrammetry and dense image matching. IEEE Conference on computer vision and pattern recognition (CVPR). San Diego, CA, USA, pp 807–814
Hirschmüller H (2008) Stereo processing by semiglobal matching and mutual information. IEEE Trans Pattern Anal Mach Intell 30(2):328–341
Hobrough G (1959) Automatic stereoplotting. In: Photogrammetric engineering and remote sensing 25.5
ifp (2016). *SURE - Photogrammetric surface reconstruction from imagery.* Accessed: 2018-05-17. http://www.ifp.uni-stuttgart.de/publications/software/sure/index.en.html
IGN-ENSG (2018). *Micmac.* Accessed: 2018-05-16. https://micmac.ensg.eu/index.php/Accueil
ISTI-CNR, (2016) Meshlab. Accessed: 2018–05-21.http://www.meshlab.net
Jähne B (2005) In: Digital image processing, 6 revised and extended edn. Springer, Berlin
Jazayeri I, Fraser CS (2010) Interest operators for featured-based matching in close range photogrammetry. Photogramm Rec 25(129):24–41
Kazhdan, M. and H. Hoppe (2013). "Screened Poisson surface reconstruction". In: *ACM Transactions on graphics* 32.3, 29:1–29:13. https://doi.org/10.1145/2487228.2487237
Kazhdan M, Bolitho M, Hoppe H (2006) Poisson surface reconstruction. In: Polthier K, Sheffer A (eds) Eurographics symposium on geometry processing. Wichmann, Berlin/Offenbach, Germany, pp 59–70
Lingua A, Marenchino D, Nex F (2009) Performance analysis of the sift operator for automatic feature extraction and matching in photogrammetric applications. Sensors 9:3745–3766. https://doi.org/10.3390/s90503745
Lowe D (2004) Interest operators for featured-based matching in close range photogrammetry. Int J Comput Vis 60(2):91–110
Moravec HP (1977) Towards automatic visual obstacle avoidance. http://www.frc.ri.cmu.edu/~hpm/project.archive/robot.papers/1977/aips.txt, Accessed by 11 May 2018

Moravec HP (1979) Visual mapping by a robot rover. http://www.frc.ri.cmu.edu/~hpm/project.archive/robot.papers/1979/ij79.txt, Accessed by 11 May 2018
nFrames (2018). *nframes*. Accessed: 2018-05-17.http://www.nframes.com/
Niederheiser, R. et al. (2016). "Deriving 3D point clouds from terrestrial photographs - Comparison of different sensors and software". In: *The International Archives of the Photogrammetry, Remote Sensing and Spatial Information Sciences*, Vol. XLI-B5. https://doi.org/10.5194/isprsarchives-XLI-B5-685-2016
Pierrot-Deseilligny, M. and N. Paparoditis (2006). "A multiresolution and optimization-based image matching approach: An application to surface reconstruction from spot5-hrs stereo imagery". In: *The International Archives of the Photogrammetry, Remote Sensing and Spatial Information Sciences*, Vol. XXXVI-1/W41. Ankara, Turkey
Pix4D (2019). *Pix4d Mapper.* Accessed: 2019-08-31. https://www.pix4d.com/
PMVS (2008). *Patch-based multi-view stereo software (PMVS).*Accessed: 2018-05-22.https://www.di.ens.fr/pmvs/pmvs-1/index.html
Pratt WK (2014) Introduction to digital image processing. CRC Press, Florida
Reiterer A, Huber N, Bauer A (2010) Image-based detection and matching of homologue points using feature-vectors—functionality and evaluation in a deformation measurement system. In: XXXVII, pp 510–515
Remondino F, et al (2013) Dense image matching: comparisons and analyses. In: Digital heritage international congress (DigitalHeritage), 2013. https://doi.org/10.1109/DigitalHeritage.2013.6743712
Rodehorst V, Koschan A (2006) Comparison and evaluation of feature point detections. In: Gruendig L, Altan, MO (eds) 5th Turkish-German joint geodetic days. Berlin, Germany
Rosenholm D (1987) Least squares matching method: some experimental results. Photogramm Rec 12(70):493–512
Rosten E, Drummond T (2006) Machine learning for high-speed corner detection. In: ECCV'06 Proceedings of the 9th European conference on computer vision, vol I, Gratz, Austria, pp 430–443. https://doi.org/10.1007/11744023_34
Rothermel, M. et al. (2012). "SURE: photogrammetric surfacereconstruction from imagery". In: *Proceedings LC3D Workshop*.Berlin, Germany
Rupnik, E., M. Daakir, and M. P. Pierrot-Deseilligny (2017). "Amulti-view dense image matching method for high-resolution aerialimagery based on a graph network".In: *Open Geospatial Data, Software and Standards* 2.14. https://doi.org/10.1186/s40965-017-0027-2
Shannon C (1948) A mathematical theory of communication. Bell Syst Tech J 27(3):379–423. https://doi.org/10.1002/j.1538-7305.1948.tb01338.x
Smith SM, Brady JM (1997) Susan–a new approach to low level image processing. Int J Comput Vis 23(1):45–78
Stylianidis E (2003) A new digital interest point operator for close range photogrammetry. In: The international archives of the photogrammetry, remote sensing and spatial information sciences, vol 34.5/W12. Corfu, Greece, pp 319–324
Sun Y et al (2014) L2-sift: Sift feature extraction and matching for large images in large-scale aerial photogrammetry. ISPRS J Photogramm Remote Sens 91:1–16. https://doi.org/10.1016/j.isprsjprs.2014.02.001
Svensk J (2017) Evaluation of aerial image stereo matching methods for forest variable estimation. MSc, Linköping University, Linköping, Sweden
Szeliski R (2011) Computer vision—algorithms and applications. Springer, Berlin. https://doi.org/10.1007/978-1-84882-935-0
VisualSFM (2018). *VisualSFM: a visual structure from motion system.* Accessed: 2018-05-22. http://ccwu.me/vsfm/
Wallis RH (1976) An approach for the space variant restoration and enhancement of images. In: Symposium on current mathematical problems in image science. Monterey, California, USA
Wenzel K et al (2013) SURE - The ifp software for dense image matching. In: Fritsch Dieter (ed) Photogrammetric Week '13. Wichmann, Berlin/Offenbach, Germany, pp 59–70

Yan L, et al (2016) A multi-view dense image matching method for high-resolution aerial imagery based on a graph network. In: Remote sensing 8.10. https://doi.org/10.3390/re8100799
Zheltov SY, Sibiryakov, AV (1997) Optical 3-D measurement techniques IV. In: Grün A, Kahmen H (eds) Chappter Adaptive subpixel cross-correlation in a point correspondence problem. Wichmann Verlag, Heidelberg, pp 86–95

Appendix A
The Venice Charter 1964

International Charter for the Conservation and Restoration of Monuments and Sites (The Venice Charter 1964)

E. Stylianidis, *Photogrammetric Survey for the Recording and Documentation of Historic Buildings*, Springer Tracts in Civil Engineering,
https://doi.org/10.1007/978-3-030-47310-5

CONSEIL INTERNATIONAL DES MONUMENTS ET DES SITES ICOMOS INTERNATIONAL COUNCIL ON MONUMENTS AND SITES

INTERNATIONAL CHARTER FOR THE CONSERVATION AND RESTORATION OF MONUMENTS AND SITES (THE VENICE CHARTER 1964)

IInd International Congress of Architects and Technicians of Historic Monuments, Venice, 1964.

Adopted by ICOMOS in 1965.

Imbued with a message from the past, the historic monuments of generations of people remain to the present day as living witnesses of their age-old traditions. People are becoming more and more conscious of the unity of human values and regard ancient monuments as a common heritage. The common responsibility to safeguard them for future generations is recognized. It is our duty to hand them on in the full richness of their authenticity.

It is essential that the principles guiding the preservation and restoration of ancient buildings should be agreed and be laid down on an international basis, with each country being responsible for applying the plan within the framework of its own culture and traditions.

By defining these basic principles for the first time, the Athens Charter of 1931 contributed towards the development of an extensive international movement which has assumed concrete form in national documents, in the work of ICOM and UNESCO and in the establishment by the latter of the International Centre for the Study of the Preservation and the Restoration of Cultural Property. Increasing awareness and critical study have been brought to bear on problems which have continually become more complex and varied; now the time has come to examine the Charter afresh in order to make a thorough study of the principles involved and to enlarge its scope in a new document.

Accordingly, the IInd International Congress of Architects and Technicians of Historic Monuments, which met in Venice from May 25th to 31st 1964, approved the following text:

DEFINITIONS

Article 1.

The concept of a historic monument embraces not only the single architectural work but also the urban or rural setting in which is found the evidence of a particular civilization, a significant development or a historic event. This applies not only to great works of art but also to more modest works of the past which have acquired cultural significance with the passing of time.

Article 2.

The conservation and restoration of monuments must have recourse to all the sciences and techniques which can contribute to the study and safeguarding of the architectural heritage.

Article 3.

The intention in conserving and restoring monuments is to safeguard them no less as works of art than as historical evidence.

CONSERVATION

Article 4.

It is essential to the conservation of monuments that they be maintained on a permanent basis.

Article 5.

The conservation of monuments is always facilitated by making use of them for some socially useful purpose. Such use is therefore desirable but it must not change the lay-out or decoration of the building. It is within these limits only that modifications demanded by a change of function should be envisaged and may be permitted.

Article 6.

The conservation of a monument implies preserving a setting which is not out of scale. Wherever the traditional setting exists, it must be kept. No new construction, demolition or modification which would alter the relations of mass and colour must be allowed.

Article 7.

A monument is inseparable from the history to which it bears witness and from the setting in which it occurs. The moving of all or part of a monument cannot be allowed except where the safeguarding of that monument demands it or where it is justified by national or international interest of paramount importance.

Article 8.

Items of sculpture, painting or decoration which form an integral part of a monument may only be removed from it if this is the sole means of ensuring their preservation.

RESTORATION

Article 9.

The process of restoration is a highly specialized operation. Its aim is to preserve and reveal the aesthetic and historic value of the monument and is based on respect for original material and authentic documents. It must stop at the point where conjecture begins, and in this case moreover any extra work which is indispensable must be distinct from the architectural composition and must bear a contemporary stamp. The restoration in any case must be preceded and followed by an archaeological and historical study of the monument.

Article 10.

Where traditional techniques prove inadequate, the consolidation of a monument can be achieved by the use of any modern technique for conservation and construction, the efficacy of which has been shown by scientific data and proved by experience.

Article 11.

The valid contributions of all periods to the building of a monument must be respected, since unity of style is not the aim of a restoration. When a building includes the superimposed work of different periods, the revealing of the underlying state can only be justified in exceptional circumstances and when what is removed is of little interest and the material which is brought to light is of great historical, archaeological or aesthetic value, and its state of preservation good enough to justify the action. Evaluation of the importance of the elements involved and the decision as to what may be destroyed cannot rest solely on the individual in charge of the work.

Article 12.

Replacements of missing parts must integrate harmoniously with the whole, but at the same time must be distinguishable from the original so that restoration does not falsify the artistic or historic evidence.

Article 13.

Additions cannot be allowed except in so far as they do not detract from the interesting parts of the building, its traditional setting, the balance of its composition and its relation with its surroundings.

HISTORIC SITES

Article 14.

The sites of monuments must be the object of special care in order to safeguard their integrity and ensure that they are cleared and presented in a seemly manner. The work of conservation and restoration carried out in such places should be inspired by the principles set forth in the foregoing articles.

EXCAVATIONS

Article 15.

Excavations should be carried out in accordance with scientific standards and the recommendation defining international principles to be applied in the case of archaeological excavation adopted by UNESCO in 1956.

Ruins must be maintained and measures necessary for the permanent conservation and protection of architectural features and of objects discovered must be taken. Furthermore, every means must be taken to facilitate the understanding of the monument and to reveal it without ever distorting its meaning.

All reconstruction work should however be ruled out "*a priori*". Only anastylosis, that is to say, the reassembling of existing but dismembered parts can be permitted. The material used for integration should always be recognizable and its use should be the least that will ensure the conservation of a monument and the reinstatement of its form.

PUBLICATION

Article 16.

In all works of preservation, restoration or excavation, there should always be precise documentation in the form of analytical and critical reports, illustrated with drawings and photographs. Every stage of the work of clearing, consolidation, rearrangement and integration, as well as technical and formal features identified during the course of the work, should be included. This record should be placed in the archives of a public institution and made available to research workers. It is recommended that the report should be published.

The following persons took part in the work of the Committee for drafting the International Charter for the Conservation and Restoration of Monuments:

Piero Gazzola (Italy), Chairman
Raymond Lemaire (Belgium), Reporter
José Bassegoda-Nonell (Spain)
Luis Benavente (Portugal)
Djurdje Boskovic (Yugoslavia)
Hiroshi Daifuku (UNESCO)
P.L. de Vrieze (Netherlands)
Harald Langberg (Denmark)
Mario Matteucci (Italy)
Jean Merlet (France)
Carlos Flores Marini (Mexico)
Roberto Pane (Italy)
S.C.J. Pavel (Czechoslovakia)
Paul Philippot (ICCROM)
Victor Pimentel (Peru)
Harold Plenderleith (ICCROM)
Deoclecio Redig de Campos (Vatican)
Jean Sonnier (France)
Francois Sorlin (France)
Eustathios Stikas (Greece)
Gertrud Tripp (Austria)
Jan Zachwatovicz (Poland)
Mustafa S. Zbiss (Tunisia)

Appendix B
World Heritage Convention 1972

UNESCO Convention Concerning the Protection of the World Cultural and Natural Heritage 1972

E. Stylianidis, *Photogrammetric Survey for the Recording and Documentation of Historic Buildings*, Springer Tracts in Civil Engineering,
https://doi.org/10.1007/978-3-030-47310-5

UNITED NATIONS EDUCATIONAL, SCIENTIFIC AND CULTURAL ORGANISATION

CONVENTION CONCERNING THE PROTECTION OF THE WORLD CULTURAL AND NATURAL HERITAGE

Adopted by the General Conference at its seventeenth session
Paris, 16 november 1972

English Text

CONVENTION CONCERNING THE PROTECTION OF THE WORLD CULTURAL AND NATURAL HERITAGE

The General Conference of the United Nations Educational, Scientific and Cultural Organization meeting in Paris from 17 October to 21 November 1972, at its seventeenth session,

Noting that the cultural heritage and the natural heritage are increasingly threatened with destruction not only by the traditional causes of decay, but also by changing social and economic conditions which aggravate the situation with even more formidable phenomena of damage or destruction,

Considering that deterioration or disappearance of any item of the cultural or natural heritage constitutes a harmful impoverishment of the heritage of all the nations of the world,

Considering that protection of this heritage at the national level often remains incomplete because of the scale of the resources which it requires and of the insufficient economic, scientific, and technological resources of the country where the property to be protected is situated,

Recalling that the Constitution of the Organization provides that it will maintain, increase, and diffuse knowledge by assuring the conservation and protection of the world's heritage, and recommending to the nations concerned the necessary international conventions,

Considering that the existing international conventions, recommendations and resolutions concerning cultural and natural property demonstrate the importance, for all the peoples of the world, of safeguarding this unique and irreplaceable property, to whatever people it may belong,

Considering that parts of the cultural or natural heritage are of outstanding interest and therefore need to be preserved as part of the world heritage of mankind as a whole,

Considering that, in view of the magnitude and gravity of the new dangers threatening them, it is incumbent on the international community as a whole to participate in the protection of the cultural and natural heritage of outstanding universal value, by the granting of collective assistance which, although not taking the place of action by the State concerned, will serve as an efficient complement thereto,

Considering that it is essential for this purpose to adopt new provisions in the form of a convention establishing an effective system of collective protection of the cultural and natural heritage of outstanding universal value, organized on a permanent basis and in accordance with modern scientific methods,

Having decided, at its sixteenth session, that this question should be made the subject of an international convention,

Adopts this sixteenth day of November 1972 this Convention.

I. DEFINITION OF THE CULTURAL AND NATURAL HERITAGE

Article 1

For the purpose of this Convention, the following shall be considered as "cultural heritage":

monuments: architectural works, works of monumental sculpture and painting, elements or structures of an archaeological nature, inscriptions, cave dwellings and combinations of features, which are of outstanding universal value from the point of view of history, art or science;

groups of buildings: groups of separate or connected buildings which, because of their architecture, their homogeneity or their place in the landscape, are of outstanding universal value from the point of view of history, art or science;

sites: works of man or the combined works of nature and man, and areas including archaeological sites which are of outstanding universal value from the historical, aesthetic, ethnological or anthropological point of view.

Article 2

For the purposes of this Convention, the following shall be considered as "natural heritage":

natural features consisting of physical and biological formations or groups of such formations, which are of outstanding universal value from the aesthetic or scientific point of view;

geological and physiographical formations and precisely delineated areas which constitute the habitat of threatened species of animals and plants of outstanding universal value from the point of view of science or conservation;

natural sites or precisely delineated natural areas of outstanding universal value from the point of view of science, conservation or natural beauty.

Article 3

It is for each State Party to this Convention to identify and delineate the different properties situated on its territory mentioned in Articles 1 and 2 above.

II. NATIONAL PROTECTION AND INTERNATIONAL PROTECTION OF THE CULTURAL AND NATURAL HERITAGE

Article 4

Each State Party to this Convention recognizes that the duty of ensuring the identification, protection, conservation, presentation and transmission to future generations of the cultural and natural heritage referred to in Articles 1 and 2 and situated on its territory, belongs primarily to that State. It will do all it can to this end, to the utmost of its own resources and, where appropriate, with any international assistance and co-operation, in particular, financial, artistic, scientific and technical, which it may be able to obtain.

Article 5

To ensure that effective and active measures are taken for the protection, conservation and presentation of the cultural and natural heritage situated on its territory, each State Party to this Convention shall endeavor, in so far as possible, and as appropriate for each country:

(a) to adopt a general policy which aims to give the cultural and natural heritage a function in the life of the community and to integrate the protection of that heritage into comprehensive planning programmes;

(b) to set up within its territories, where such services do not exist, one or more services for the protection, conservation and presentation of the cultural and natural heritage with an appropriate staff and possessing the means to discharge their functions;

(c) to develop scientific and technical studies and research and to work out such operating methods as will make the State capable of counteracting the dangers that threaten its cultural or natural heritage;

(d) to take the appropriate legal, scientific, technical, administrative and financial measures necessary for the identification, protection, conservation, presentation and rehabilitation of this heritage; and

(e) to foster the establishment or development of national or regional centres for training in the protection, conservation and presentation of the cultural and natural heritage and to encourage scientific research in this field.

Article 6

1. Whilst fully respecting the sovereignty of the States on whose territory the cultural and natural heritage mentioned in Articles 1 and 2 is situated, and without prejudice to property right provided by national legislation, the States Parties to this Convention recognize that such heritage constitutes a world heritage for whose protection it is the duty of the international community as a whole to co-operate.
2. The States Parties undertake, in accordance with the provisions of this Convention, to give their help in the identification, protection, conservation and presentation of the cultural and natural heritage referred to in paragraphs 2 and 4 of Article 11 if the States on whose territory it is situated so request.
3. Each State Party to this Convention undertakes not to take any deliberate measures which might damage directly or indirectly the cultural and natural heritage referred to in Articles 1 and 2 situated on the territory of other States Parties to this Convention.

Article 7

For the purpose of this Convention, international protection of the world cultural and natural heritage shall be understood to mean the establishment of a system of international co-operation and assistance designed to support States Parties to the Convention in their efforts to conserve and identify that heritage.

III INTERGOVERNMENTAL COMMITTEE FOR THE PROTECTION OF THE WORLD CULTURAL AND NATURAL HERITAGE

Article 8

1. An Intergovernmental Committee for the Protection of the Cultural and Natural Heritage of Outstanding Universal Value, called "the World Heritage Committee", is hereby established within the United Nations Educational, Scientific and Cultural Organization. It shall be composed of 15 States Parties to the Convention, elected by States Parties to the Convention meeting in general assembly during the ordinary session of the General Conference of the United Nations Educational, Scientific and Cultural Organization. The number of States members of the Committee shall be increased to 21 as from the date of the ordinary session of the General Conference following the entry into force of this Convention for at least 40 States.

2. Election of members of the Committee shall ensure an equitable representation of the different regions and cultures of the world.

3. A representative of the International Centre for the Study of the Preservation and Restoration of Cultural Property (Rome Centre), a representative of the International Council of Monuments and Sites (ICOMOS) and a representative of the International Union for Conservation of Nature and Natural Resources (IUCN), to whom may be added, at the request of States Parties to the Convention meeting in general assembly during the ordinary sessions of the General Conference of the United Nations Educational, Scientific and Cultural Organization, representatives of other intergovernmental or non-governmental organizations, with similar objectives, may attend the meetings of the Committee in an advisory capacity.

Article 9

1. The term of office of States members of the World Heritage Committee shall extend from the end of the ordinary session of the General Conference during which they are elected until the end of its third subsequent ordinary session.

2. The term of office of one-third of the members designated at the time of the first election shall, however, cease at the end of the first ordinary session of the General Conference following that at which they were elected; and the term of office of a further third of the members designated at the same time shall cease at the end of the second ordinary session of the General Conference following that at which they were elected. The names of these members shall be chosen by lot by the President of the General Conference of the United Nations Educational, Scientific and Cultural Organization after the first election.

3. States members of the Committee shall choose as their representatives persons qualified in the field of the cultural or natural heritage.

Article 10

1. The World Heritage Committee shall adopt its Rules of Procedure.

2. The Committee may at any time invite public or private organizations or individuals to participate in its meetings for consultation on particular problems.

3. The Committee may create such consultative bodies as it deems necessary for the performance of its functions.

Article 11

1. Every State Party to this Convention shall, in so far as possible, submit to the World Heritage Committee an inventory of property forming part of the cultural and natural heritage, situated in its territory and suitable for inclusion in the list provided for in paragraph 2 of this Article. This inventory, which shall not be considered exhaustive, shall include documentation about the location of the property in question and its significance.

2. On the basis of the inventories submitted by States in accordance with paragraph 1, the Committee shall establish, keep up to date and publish, under the title of "World Heritage List," a list of properties forming part of the cultural heritage and natural heritage, as defined in Articles 1 and 2 of this Convention, which it considers as having outstanding universal value in terms of such criteria as it shall have established. An updated list shall be distributed at least every two years.

3. The inclusion of a property in the World Heritage List requires the consent of the State concerned. The inclusion of a property situated in a territory, sovereignty or jurisdiction over which is claimed by more than one State shall in no way prejudice the rights of the parties to the dispute.

4. The Committee shall establish, keep up to date and publish, whenever circumstances shall so require, under the title of "list of World Heritage in Danger", a list of the property appearing in the World Heritage List for the conservation of which major operations are necessary and for which assistance has been requested under this Convention. This list shall contain an estimate of the cost of such operations. The list may include only such property forming part of the cultural and natural heritage as is threatened by serious and specific dangers, such as the threat of disappearance caused by accelerated deterioration, large-scale public or private projects or rapid urban or tourist development projects; destruction caused by changes in the use or ownership of the land; major alterations due to unknown causes; abandonment for any reason whatsoever; the outbreak or the threat of an armed conflict; calamities and cataclysms; serious fires, earthquakes, landslides; volcanic eruptions; changes in water level, floods and tidal waves. The Committee may at any time, in case of urgent need, make a new entry in the List of World Heritage in Danger and publicize such entry immediately.

5. The Committee shall define the criteria on the basis of which a property belonging to the cultural or natural heritage may be included in either of the lists mentioned in paragraphs 2 and 4 of this article.

6. Before refusing a request for inclusion in one of the two lists mentioned in paragraphs 2 and 4 of this article, the Committee shall consult the State Party in whose territory the cultural or natural property in question is situated.

7. The Committee shall, with the agreement of the States concerned, co-ordinate and encourage the studies and research needed for the drawing up of the lists referred to in paragraphs 2 and 4 of this article.

Article 12

The fact that a property belonging to the cultural or natural heritage has not been included in either of the two lists mentioned in paragraphs 2 and 4 of Article 11 shall in no way be construed to mean that it does not have an outstanding universal value for purposes other than those resulting from inclusion in these lists.

Article 13

1. The World Heritage Committee shall receive and study requests for international assistance formulated by States Parties to this Convention with respect to property forming part of the cultural or natural heritage, situated in their territories, and included or potentially suitable for inclusion in the lists mentioned referred to in paragraphs 2 and 4 of Article 11. The purpose of such requests may be to secure the protection, conservation, presentation or rehabilitation of such property.

2. Requests for international assistance under paragraph 1 of this article may also be concerned with identification of cultural or natural property defined in Articles 1 and 2, when preliminary investigations have shown that further inquiries would be justified.

3. The Committee shall decide on the action to be taken with regard to these requests, determine where appropriate, the nature and extent of its assistance, and authorize the conclusion, on its behalf, of the necessary arrangements with the government concerned.

4. The Committee shall determine an order of priorities for its operations. It shall in so doing bear in mind the respective importance for the world cultural and natural heritage of the property requiring protection, the need to give international assistance to the property most representative of a natural environment or of the genius and the history of the peoples of the world, the urgency of the work to be done, the resources available to the States on whose territory the threatened property is situated and in particular the extent to which they are able to safeguard such property by their own means.

5. The Committee shall draw up, keep up to date and publicize a list of property for which international assistance has been granted.

6. The Committee shall decide on the use of the resources of the Fund established under Article 15 of this Convention. It shall seek ways of increasing these resources and shall take all useful steps to this end.

7. The Committee shall co-operate with international and national governmental and non-governmental organizations having objectives similar to those of this Convention. For the implementation of its programmes and projects, the Committee may call on such organizations, particularly the International Centre for the Study of the Preservation and Restoration of Cultural Property (the Rome Centre), the International Council of Monuments and Sites (ICOMOS) and the International Union for Conservation of Nature and Natural Resources (IUCN), as well as on public and private bodies and individuals.

8. Decisions of the Committee shall be taken by a majority of two-thirds of its members present and voting. A majority of the members of the Committee shall constitute a quorum.

Article 14

1. The World Heritage Committee shall be assisted by a Secretariat appointed by the Director-General of the United Nations Educational, Scientific and Cultural Organization.

2. The Director-General of the United Nations Educational, Scientific and Cultural Organization, utilizing to the fullest extent possible the services of the International Centre for the Study of the Preservation and the Restoration of Cultural Property (the Rome Centre), the International Council of Monuments and Sites (ICOMOS) and the International Union for Conservation of Nature and Natural Resources (IUCN) in their respective areas of competence and capability, shall prepare the Committee's documentation and the agenda of its meetings and shall have the responsibility for the implementation of its decisions.

IV FUND FOR THE PROTECTION OF THE WORLD CULTURAL AND NATURAL HERITAGE

Article 15

1. A Fund for the Protection of the World Cultural and Natural Heritage of Outstanding Universal Value, called "the World Heritage Fund", is hereby established.

2. The Fund shall constitute a trust fund, in conformity with the provisions of the Financial Regulations of the United Nations Educational, Scientific and Cultural Organization.

3. The resources of the Fund shall consist of:

(a) compulsory and voluntary contributions made by States Parties to this Convention,

(b) Contributions, gifts or bequests which may be made by:

(i) other States;

(ii) the United Nations Educational, Scientific and Cultural Organization, other organizations of the United Nations system, particularly the United Nations Development Programme or other intergovernmental organizations;

(iii) public or private bodies or individuals;

(c) any interest due on the resources of the Fund;

(d) funds raised by collections and receipts from events organized for the benefit of the fund; and

(e) all other resources authorized by the Fund's regulations, as drawn up by the World Heritage Committee.

4. Contributions to the Fund and other forms of assistance made available to the Committee may be used only for such purposes as the Committee shall define. The Committee may accept contributions to be used only for a certain programme or project, provided that the Committee shall have decided on the implementation of such programme or project. No political conditions may be attached to contributions made to the Fund.

Article 16

1. Without prejudice to any supplementary voluntary contribution, the States Parties to this Convention undertake to pay regularly, every two years, to the World Heritage Fund, contributions, the amount of which, in the form of a uniform percentage applicable to all States, shall be determined by the General Assembly of States Parties to the Convention, meeting during the sessions of the General Conference of the United Nations Educational, Scientific and Cultural Organization. This decision of the General Assembly requires the majority of the States Parties present and voting, which have not made the declaration referred to in paragraph 2 of this Article. In no case shall the compulsory contribution of States Parties to the Convention exceed 1% of the contribution to the regular budget of the United Nations Educational, Scientific and Cultural Organization.

2. However, each State referred to in Article 31 or in Article 32 of this Convention may declare, at the time of the deposit of its instrument of ratification, acceptance or accession, that it shall not be bound by the provisions of paragraph 1 of this Article.

3. A State Party to the Convention which has made the declaration referred to in paragraph 2 of this Article may at any time withdraw the said declaration by notifying the Director-General of the United Nations Educational, Scientific and Cultural Organization. However, the withdrawal of the declaration shall not take effect in regard to the compulsory contribution due by the State until the date of the subsequent General Assembly of States parties to the Convention.

4. In order that the Committee may be able to plan its operations effectively, the contributions of States Parties to this Convention which have made the declaration referred to in paragraph 2 of this Article, shall be paid on a regular basis, at least every two years, and should not be less than the contributions which they should have paid if they had been bound by the provisions of paragraph 1 of this Article.

5. Any State Party to the Convention which is in arrears with the payment of its compulsory or voluntary contribution for the current year and the calendar year immediately preceding it shall not be eligible as a Member of the World Heritage Committee, although this provision shall not apply to the first election.

   The terms of office of any such State which is already a member of the Committee shall terminate at the time of the elections provided for in Article 8, paragraph 1 of this Convention.

Article 17

The States Parties to this Convention shall consider or encourage the establishment of national public and private foundations or associations whose purpose is to invite donations for the protection of the cultural and natural heritage as defined in Articles 1 and 2 of this Convention.

Article 18

The States Parties to this Convention shall give their assistance to international fund-raising campaigns organized for the World Heritage Fund under the auspices of the United Nations Educational, Scientific and Cultural Organization. They shall facilitate collections made by the bodies mentioned in paragraph 3 of Article 15 for this purpose.

V. CONDITIONS AND ARRANGEMENTS FOR INTERNATIONAL ASSISTANCE

Article 19

Any State Party to this Convention may request international assistance for property forming part of the cultural or natural heritage of outstanding universal value situated within its territory. It shall submit with its request such information and documentation provided for in Article 21 as it has in its possession and as will enable the Committee to come to a decision.

Article 20

Subject to the provisions of paragraph 2 of Article 13, sub-paragraph (c) of Article 22 and Article 23, international assistance provided for by this Convention may be granted only to property forming part of the cultural and natural heritage which the World Heritage Committee has decided, or may decide, to enter in one of the lists mentioned in paragraphs 2 and 4 of Article 11.

Article 21

1. The World Heritage Committee shall define the procedure by which requests to it for international assistance shall be considered and shall specify the content of the request, which should define the operation contemplated, the work that is necessary, the expected cost thereof, the degree of urgency and the reasons why the resources of the State requesting assistance do not allow it to meet all the expenses. Such requests must be supported by experts' reports whenever possible.

2. Requests based upon disasters or natural calamities should, by reasons of the urgent work which they may involve, be given immediate, priority consideration by the Committee, which should have a reserve fund at its disposal against such contingencies.

3. Before coming to a decision, the Committee shall carry out such studies and consultations as it deems necessary.

Article 22

Assistance granted by the World Heritage Fund may take the following forms:

(a) studies concerning the artistic, scientific and technical problems raised by the protection, conservation, presentation and rehabilitation of the cultural and natural heritage, as defined in paragraphs 2 and 4 of Article 11 of this Convention;

(b) provisions of experts, technicians and skilled labour to ensure that the approved work is correctly carried out;

(c) training of staff and specialists at all levels in the field of identification, protection, conservation, presentation and rehabilitation of the cultural and natural heritage;

(d) supply of equipment which the State concerned does not possess or is not in a position to acquire;

(e) low-interest or interest-free loans which might be repayable on a long-term basis;

(f) the granting, in exceptional cases and for special reasons, of non-repayable subsidies.

Article 23

The World Heritage Committee may also provide international assistance to national or regional centres for the training of staff and specialists at all levels in the field of identification, protection, conservation, presentation and rehabilitation of the cultural and natural heritage.

Article 24

International assistance on a large scale shall be preceded by detailed scientific, economic and technical studies. These studies shall draw upon the most advanced techniques for the protection, conservation, presentation and rehabilitation of the natural and cultural heritage and shall be consistent with the objectives of this Convention. The studies shall also seek means of making rational use of the resources available in the State concerned.

Article 25

As a general rule, only part of the cost of work necessary shall be borne by the international community. The contribution of the State benefiting from international assistance shall constitute a substantial share of the resources devoted to each programme or project, unless its resources do not permit this.

Article 26

The World Heritage Committee and the recipient State shall define in the agreement they conclude the conditions in which a programme or project for which international assistance under the terms of this Convention is provided, shall be carried out. It shall be the responsibility of the State receiving such international assistance to continue to protect, conserve and present the property so safeguarded, in observance of the conditions laid down by the agreement.

VI. EDUCATIONAL PROGRAMMES

Article 27

1. The States Parties to this Convention shall endeavor by all appropriate means, and in particular by educational and information programmes, to strengthen appreciation and respect by their peoples of the cultural and natural heritage defined in Articles 1 and 2 of the Convention.

2. They shall undertake to keep the public broadly informed of the dangers threatening this heritage and of the activities carried on in pursuance of this Convention.

Article 28

States Parties to this Convention which receive international assistance under the Convention shall take appropriate measures to make known the importance of the property for which assistance has been received and the role played by such assistance.

VII. REPORTS

Article 29

1. The States Parties to this Convention shall, in the reports which they submit to the General Conference of the United Nations Educational, Scientific and Cultural Organization on dates and in a manner to be determined by it, give information on the legislative and administrative provisions which they have adopted and other action which they have taken for the application of this Convention, together with details of the experience acquired in this field.

2. These reports shall be brought to the attention of the World Heritage Committee.

3. The Committee shall submit a report on its activities at each of the ordinary sessions of the General Conference of the United Nations Educational, Scientific and Cultural Organization.

VIII FINAL CLAUSES

Article 30

This Convention is drawn up in Arabic, English, French, Russian and Spanish, the five texts being equally authoritative.

Article 31

1. This Convention shall be subject to ratification or acceptance by States members of the United Nations Educational, Scientific and Cultural Organization in accordance with their respective constitutional procedures.

2. The instruments of ratification or acceptance shall be deposited with the Director-General of the United Nations Educational, Scientific and Cultural Organization.

Article 32

1. This Convention shall be open to accession by all States not members of the United Nations Educational, Scientific and Cultural Organization which are invited by the General Conference of the Organization to accede to it.

2. Accession shall be effected by the deposit of an instrument of accession with the Director-General of the United Nations Educational, Scientific and Cultural Organization.

Article 33

This Convention shall enter into force three months after the date of the deposit of the twentieth instrument of ratification, acceptance or accession, but only with respect to those States which have deposited their respective instruments of ratification, acceptance or accession on or before that date. It shall enter into force with respect to any other State three months after the deposit of its instrument of ratification, acceptance or accession.

Article 34

The following provisions shall apply to those States Parties to this Convention which have a federal or non-unitary constitutional system:

(a) with regard to the provisions of this Convention, the implementation of which comes under the legal jurisdiction of the federal or central legislative power, the obligations of the federal or central government shall be the same as for those States parties which are not federal States;

(b) with regard to the provisions of this Convention, the implementation of which comes under the legal jurisdiction of individual constituent States, countries, provinces or cantons that are not obliged by the constitutional system of the federation to take legislative measures, the federal government shall inform the competent authorities of such States, countries, provinces or cantons of the said provisions, with its recommendation for their adoption.

Article 35

1. Each State Party to this Convention may denounce the Convention.

2. The denunciation shall be notified by an instrument in writing, deposited with the Director-General of the United Nations Educational, Scientific and Cultural Organization.

3. The denunciation shall take effect twelve months after the receipt of the instrument of denunciation. It shall not affect the financial obligations of the denouncing State until the date on which the withdrawal takes effect.

Article 36

The Director-General of the United Nations Educational, Scientific and Cultural Organization shall inform the States members of the Organization, the States not members of the Organization which are referred to in Article 32, as well as the United Nations, of the deposit of all the instruments of ratification, acceptance, or accession provided for in Articles 31 and 32, and of the denunciations provided for in Article 35.

Article 37

1. This Convention may be revised by the General Conference of the United Nations Educational, Scientific and Cultural Organization. Any such revision shall, however, bind only the States which shall become Parties to the revising convention.

2. If the General Conference should adopt a new convention revising this Convention in whole or in part, then, unless the new convention otherwise provides, this Convention shall cease to be open to ratification, acceptance or accession, as from the date on which the new revising convention enters into force.

Article 38

In conformity with Article 102 of the Charter of the United Nations, this Convention shall be registered with the Secretariat of the United Nations at the request of the Director-General of the United Nations Educational, Scientific and Cultural Organization.

Done in Paris, this twenty-third day of November 1972, in two authentic copies bearing the signature of the President of the seventeenth session of the General Conference and of the Director-General of the United Nations Educational, Scientific and Cultural Organization, which shall be deposited in the archives of the United Nations Educational, Scientific and Cultural Organization, and certified true copies of which shall be delivered to all the States referred to in Articles 31 and 32 as well as to the United Nations.

Appendix C
ICOMOS Principles 1996

Principles for the Recording of Monuments, Groups of Buildings and Sites 1996

E. Stylianidis, *Photogrammetric Survey for the Recording and Documentation of Historic Buildings*, Springer Tracts in Civil Engineering,
https://doi.org/10.1007/978-3-030-47310-5

CONSEIL INTERNATIONAL DES MONUMENTS ET DES SITES **ICOMOS** INTERNATIONAL COUNCIL ON MONUMENTS AND SITES

PRINCIPLES FOR THE RECORDING OF MONUMENTS, GROUPS OF BUILDINGS AND SITES (1996)

Ratified by the 11th ICOMOS General Assembly in Sofia, October 1996.

As the cultural heritage is a unique expression of human achievement; and

As this cultural heritage is continuously at risk; and

As recording is one of the principal ways available to give meaning, understanding, definition and recognition of the values of the cultural heritage; and

As the responsibility for conserving and maintaining the cultural heritage rests not only with the owners but also with conservation specialists and the professionals, managers, politicians and administrators working at all levels of government, and with the public; and

As article 16 of the Charter of Venice requires, it is essential that responsible organisations and individuals record the nature of the cultural heritage.

The purpose of this document is therefore to set out the principal reasons, responsibilities, planning measures, contents, management and sharing considerations for the recording of the cultural heritage.

DEFINITIONS OF WORDS USED IN THIS DOCUMENT:

Cultural Heritage refers to monuments, groups of buildings and sites of heritage value, constituting the historic or built environment.

Recording is the capture of information which describes the physical configuration, condition and use of monuments, groups of buildings and sites, at points in time, and it is an essential part of the conservation process.

Records of monuments, groups of buildings and sites may include tangible as well as intangible evidence, and constitute a part of the documentation that can contribute to an understanding of the heritage and its related values.

THE REASONS FOR RECORDING

1. The recording of the cultural heritage is essential:

a) To acquire knowledge in order to advance the understanding of cultural heritage, its values and its evolution;

b) To promote the interest and involvement of the people in the preservation of the heritage through the dissemination of recorded information;

c) To permit informed management and control of construction works and of all change to the cultural heritage;

d) To ensure that the maintenance and conservation of the heritage is sensitive to its physical form, its materials, construction, and its historical and cultural significance.

2. Recording should be undertaken to an appropriate level of detail in order to:

a) Provide information for the process of identification, understanding, interpretation and presentation of the heritage, and to promote the involvement of the public;

b) Provide a permanent record of all monuments, groups of buildings and sites that are to be destroyed or altered in any way, or where at risk from natural events or human activities;

c) Provide information for administrators and planners at national, regional or local levels to make sensitive planning and development control policies and decisions;

d) Provide information upon which appropriate and sustainable use may be identified, and the effective research, management, maintenance programmes and construction works may be planned.

3. Recording of the cultural heritage should be seen as a priority, and should be undertaken especially:

a) When compiling a national, regional, or local inventory;

b) As a fully integrated part of research and conservation activity;

c) Before, during and after any works of repair, alteration, or other intervention, and when evidence of its history is revealed during such works;

d) When total or partial demolition, destruction, abandonment or relocation is contemplated, or where the heritage is at risk of damage from human or natural external forces;

e) During or following accidental or unforeseen disturbance which damages the cultural heritage;

f) When change of use or responsibility for management or control occurs.

RESPONSIBILITY FOR RECORDING

1. The commitment at the national level to conserve the heritage requires an equal commitment towards the recording process.

2. The complexity of the recording and interpretation processes requires the deployment of individuals with adequate skill, knowledge and awareness for the associated tasks. It may be necessary to initiate training programmes to achieve this.

3. Typically the recording process may involve skilled individuals working in collaboration, such as specialist heritage recorders, surveyors, conservators, architects, engineers, researchers, architectural historians, archaeologists above and below ground, and other specialist advisors.

4. All managers of cultural heritage are responsible for ensuring the adequate recording, quality and updating of the records.

PLANNING FOR RECORDING

1. Before new records are prepared, existing sources of information should be found and examined for their adequacy.

a) The type of records containing such information should be searched for in surveys, drawings, photographs, published and unpublished accounts and descriptions, and related documents pertaining to the origins and history of the building, group of buildings or site. It is important to search out recent as well as old records;

b) Existing records should be searched for in locations such as national and local public archives, in professional, institutional or private archives, inventories and collections, in libraries or museums;

c) Records should be searched for through consultation with individuals and organisations who have owned, occupied, recorded, constructed, conserved, or carried out research into or who have knowledge of the building, group of buildings or site.

2. Arising out of the analysis above, selection of the appropriate scope, level and methods of recording requires that:

a) The methods of recording and type of documentation produced should be appropriate to the nature of the heritage, the purposes of the record, the cultural context, and the funding or other resources available. Limitations of such resources may require a phased approach to recording. Such methods might include written descriptions and analyses, photographs (aerial or terrestrial), rectified photography, photogrammetry, geophysical survey, maps, measured plans, drawings and sketches, replicas or other traditional and modern technologies;

b) Recording methodologies should, wherever possible, use non-intrusive techniques, and should not cause damage to the object being recorded;

c) The rationale for the intended scope and the recording method should be clearly stated;

d) The materials used for compiling the finished record must be archivally stable.

CONTENT OF RECORDS

1. Any record should be identified by:

a) The name of the building, group of buildings or

b) A unique reference number;

c) The date of compilation of the record;

d) The name of the recording organisation;

e) Cross-references to related building records and reports, photographic, graphic, textual or bibliographic documentation, archaeological and environmental records.

2. The location and extent of the monument, group of buildings or site must be given accurately - this may be achieved by description, maps, plans or aerial photographs. In rural areas a map reference or triangulation to known points may be the only methods available. In urban areas an address or street reference may be sufficient.

3. New records should note the sources of all information not obtained directly from the monument, group of buildings or site itself.

4. Records should include some or all of the following information:

a) The type, form and dimensions of the building, monument or site;

b) The interior and exterior characteristics, as appropriate, of the monument, group of buildings or site;

c) The nature, quality, cultural, artistic and scientific significance of the heritage and its components and the cultural, artistic and scientific significance of:

- the materials, constituent parts and construction, decoration, ornament or inscriptions
- services, fittings and machinery,
- ancillary structures, the gardens, landscape and the cultural, topographical and natural features of the site;

d) The traditional and modern technology and skills used in construction and maintenance;

e) Evidence to establish the date of origin, authorship, ownership, the original design, extent, use and decoration;

f) Evidence to establish the subsequent history of its uses, associated events, structural or decorative alterations, and the impact of human or natural external forces;

g) The history of management, maintenance and repairs;

h) Representative elements or samples of construction or site materials;

i) An assessment of the current condition of the heritage;

j) An assessment of the visual and functional relationship between the heritage and its setting;

k) An assessment of the conflicts and risks from human or natural causes, and from environmental pollution or adjacent land uses.

5. In considering the different reasons for recording (see Section 1.2 above) different levels of detail will be required. All the above information, even if briefly stated, provides important data for local planning and building control and management. Information in greater detail is generally required for the site or building owner's, manager's or user's purposes for conservation, maintenance and use.

MANAGEMENT, DISSEMINATION AND SHARING OF RECORDS

1. The original records should be preserved in a safe archive, and the archive's environment must ensure permanence of the information and freedom from decay to recognised international standards.

2. A complete back-up copy of such records should be stored in a separate safe location.

3. Copies of such records should be accessible to the statutory authorities, to concerned professionals and to the public, where appropriate, for the purposes of research, development controls and other administrative and legal processes.

4. Up-dated records should be readily available, if possible on the site, for the purposes of research on the heritage, management, maintenance and disaster relief.

5. The format of the records should be standardised, and records should be indexed wherever possible to facilitate the exchange and retrieval of information at a local, national or international level.

6. The effective assembly, management and distribution of recorded information requires, wherever possible, the understanding and the appropriate use of up-to-date information technology.

7. The location of the records should be made public.

8. A report of the main results of any recording should be disseminated and published, when appropriate.

The manufacturer's authorised representative in the EU is Springer
Nature Customer Service Centre GmbH, Europaplatz 3, 69115 Heidelberg,
Germany. If you have any concerns regarding our products, please
contact ProductSafety@springernature.com

Printed and bound by CPI Group (UK) Ltd, Croydon, CR0 4YY
30/09/2026
02225497-0011